The Open Univ

Block 1

Contexts, conquests and consequences – a contemplation of engineering

This publication forms part of an Open University course T173 *Engineering the future*. Details of this and other Open University courses can be obtained from the Student Registration and Enquiry Service, The Open University, PO Box 197, Milton Keynes MK7 6BJ, United Kingdom: tel. +44 (0)845 300 60 90, email general-enquiries@open.ac.uk

Alternatively, you may visit the Open University website at http://www.open.ac.uk where you can learn more about the wide range of courses and packs offered at all levels by The Open University.

To purchase a selection of Open University course materials visit http://www.ouw.co.uk, or contact Open University Worldwide, Michael Young Building, Walton Hall, Milton Keynes MK7 6AA, United Kingdom for a brochure. tel. +44 (0)1908 858793; fax +44 (0)1908 858787; email ouw-customer-services@open.ac.uk

The Open University
Walton Hall, Milton Keynes
MK7 6AA

First published 2001. Second edition 2002. Third Edition 2007.

Edited and designed by The Open University.

Typeset by SR Nova Pvt. Ltd, Bangalore, India

Printed in the United Kingdom by Thanet Press Ltd, Margate

ISBN 978 0 7492 2345 8

3.1

Part 1
Engineering history

T173 Course Team

Academic Staff

Dr Michael Fitzpatrick (Course Team Chair)
Graham Weaver
Dr Nicholas Braithwaite
Adrian Demaid
Dr Bill Kennedy
James Moffatt
Dr George Weidmann
Mark Endean
Jim Flood
Dr Suresh Nesaratnam
Professor Bill Plumbridge

Consultant

Shelagh Lewis

Production Staff

Sylvan Bentley (Picture Research)
Philippa Broadbent (Materials Procurement)
Daphne Cross (Materials Procurement)
Tony Duggan (Project Control)
Elsie Frost (Course Team Secretary)
Andy Harding (Course Manager)
Richard Hoyle (Designer)
Allan Jones (Editor)
Howard Twiner (Graphic Artist)

Contents

1 Introduction

1.1 Engineering the future

As the title *Engineering the Future* indicates, this is a course about engineering. It covers the principles and practice of a wide range of activities that fall under the general heading of 'engineering'; for example, design and manufacturing. In doing so, the course draws heavily on many examples of the *application* of technology, in its innumerable guises.

The course title also refers to engineering *the future*. So, in looking at particular facets of engineering, we will try to look forward to future trends, and, where possible, to investigate case studies of products or methods which are yet to come to fruition.

A useful introduction to the vast field of engineering is to look at its historical development. That is the purpose of this first block of the course. Block 1 will set the narrative for the rest of the course by beginning to look towards future developments, and by introducing themes that will run through the course.

There are several components to this block of the course: this written text, a *Study Guide*, some audio tracks, and so on. You can read through this text by itself, but you should read the *Study Guide* to find out how the various elements of the block fit together. This will help you plan your study of the block as a whole. There are also references to the set book, the *Sciences Good Study Guide*, which contains information that may be helpful to you during the study of this block.

1.2 Aims of this block

The aims of this block of the course are as follows:

- To introduce the course by defining what engineering is, and the social and economic contexts of its practice.
- To reveal the origins and landmarks of engineering.
- To show engineering as a creative and intellectual human activity.
- To explore why and how we engineer.

2 First words

2.1 Possibilities and constraints

The purpose of this course is to explore *how* engineering is done. In order to gain an overall perspective of the ground you are going to explore, this introductory part will investigate a range of questions which will start us on our way.

First, of course, you will need to understand what engineering *is*. Perhaps you think you already know that, but engineering covers a tremendously broad sweep of human activity. When as a course team we tried to define it, the best we came up with was the rather smug answer that 'engineering is what engineers do!' I don't like that very much because it suggests that to do any engineering you must become a special person – an *engineer* – and be initiated into that tribe. If you have ever put up some shelves, or exercised your DIY skills to manufacture something useful from scratch, then I reckon you have done some engineering.

The degree to which you have been an engineer while doing these tasks might be considered to depend on whether you just followed some instructions or whether you supplied some creative input of your own. Did you buy a set of shelves ready to put up with diagrams of how to do it, or did you just buy the wood, some varnish, screws and brackets, and work it out for yourself? An analogy is whether you follow a recipe word for word, or use experience from your cooking repertoire to make improvements to it.

If you tackled the problem yourself from scratch, without being provided with a starting point, then in general terms you have had to create a design to meet your purpose, organize your resources, and fabricate your product. These are certainly three things that professional engineers do. In part, the difference is more to do with the scale and complexity of projects than in the approach; that is why I look at shelf-assembly as engineering. However, to be a *professional* engineer you do have to join the tribe. To be able to handle greater scale and complexity, efficiently and effectively, requires a wider range of skills which are developed through education, training and experience. Such skills as management, financial planning, mathematical aptitude and business strategy all fall within the purview of the professional engineer. This course does not cover such skills, but you will find pointers towards them and their significance in the engineering context.

Scale and complexity of planning are not all that characterize a 'professional' engineer's work. If you set out to make some shelves, you will almost certainly have a mental image of what you want; and your image will be derived from something you have already seen. When we think of the names of 'great' engineers – James Watt (1736–1819), George Stephenson (1781–1848), Isambard Kingdom Brunel (1806–59), Barnes Wallis (1887–1979), Frank Whittle (1907–1996) – their greatness is recognized because they went *beyond the frontiers which existed at their time*. They saw new ways round problems, as well as having the drive and enthusiasm to bring their ideas to fruition.

Of course, not every new idea can be made to work: Leonardo da Vinci (1452–1519) drew pictures of aircraft and submarines but did not build either.

Figure 1.1 Flight of fancy – Leonardo da Vinci's sketch for a flying machine

What makes engineering exciting is exploring what is possible. Impossibility may come from all sorts of constraints: inadequacy of materials or processes, lack of finance, even from ▼The laws of nature▲. The best engineers are those who push hardest at the constraints to move them out of the way of what they want to achieve. We shall think about this more later on.

2.2 Contexts and quantities

Engineering started in the distant past: people creating things out of what came to hand to meet some purpose. It seems to me that the ability to 'engineer' is something inherently human, so one basic question to ask is what being human means.

Evolution theory claims we have common ancestors with apes, but apes do not go to the Moon or even make crude stone tools. When did humans become distinct and what are the distinctions? Without these distinctions we could not have become engineers of either amateur or professional calibre. And somehow, through the course of our history, we have moved on from primitive beginnings to our present highly engineered world. We will briefly follow the trail of that history because it will inform us of fuller answers to 'what engineering is'.

Next it is important to understand *why* we do engineering. As we look into this we shall find that the kinds of *purpose* which set us going on an engineering track have changed with time, largely as a consequence of the engineering that has gone before. There is a lot to consider here; I do not think it is putting it too strongly to say that our present social structures derive from this question of why we do engineering. If that is so, then a related question is to ask where engineering is taking our species in the future. Not for nothing is the course entitled *Engineering the Future.*

So far my questions have been somewhat abstracted from the actual business of *doing* engineering – the 'how' that I proposed as our goal at the start of Section 2. But we need to consider also questions like: 'What do we do engineering with?' and 'How do we get better at it?' As we examine these matters you will come to appreciate that there are both external resources (materials, energy, labour, money) and internal processes (ways of thinking, organizational and communications skills) that have to come together to get beyond the set-of-shelves stage.

Those are the areas covered in this introduction to set engineering into its human, social, economic, resource and intellectual contexts. Later in the course this broad view will be less obvious as you concentrate on more specialized aspects, so it is important to scan the whole map first. On the way I shall be looking back into history and prehistory for examples, and to spot some of the key moments in our development which have brought us to our present world.

▼The laws of nature▲

In every sphere of engineering, there are always limits to what can be achieved. Practically, these may be constraints such as time or money, but fundamental physical rules can be barriers also.

There are limits to how tall a building can be constructed; what the longest span of a bridge can be; how fast information can be sent along a telephone wire; or how much energy can be stored in a battery. Exactly what governs the limit depends on the problem. My first two examples are limited by the properties of construction materials; the third is dependent on how signals propagate in the copper wires used for telephone cables; the final one is to do with how much energy can be harnessed from the chemicals inside the battery.

We shall talk more about such fundamental rules in Block 3. However, you should be aware that constraints do exist which cannot be removed with more money. Sometimes a whole new design philosophy is needed to solve a problem: fibre optic cables, for example, allow far more information to be sent per second over long distances than a traditional pair of metal wires would.

There is one more theme to run through this block. Engineering is a quantitative art. At the very least, at the design stage of a product, decisions about size have to be made. Even for this book, decisions were needed as to how big the pages were to be, what the thickness and opacity of the paper should be, and how big the type should be. When constructing a bridge, the width of a river must be known in order to specify parts for it. Beyond that, there need to be estimates of money, time, labour and materials. Indeed, the whole gamut of materials properties has been measured and tabulated as understanding of the physical world has developed.

To talk quantities you need *numbers* and *units of measure*. This course will assume you are numerate, and so it will not teach mathematics directly, but you need to become entirely familiar with the units of measurement of such physical quantities as length, time, mass and power.

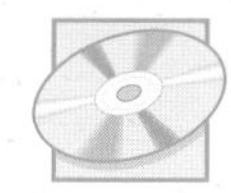

Remember that Maths Help and numeracy practice are available from the *Sciences Good Study Guide* and the CD-ROM.

We shall use the *Système International* (SI) of units, which is an international standard form of the metric system. Scattered throughout this block are sections which gradually introduce and develop this unit system, which you will need continuously in later parts of your study.

2.3 Summary

Five questions to be explored in this block are:

1 Who *are* the engineers?

2 What *is* engineering?

3 Why do we do engineering?

4 What do we *use* to do engineering?

5 Where is engineering taking us?

3 Who are the engineers?

3.1 What makes humans different?

To start ourselves on the engineering trail, we should look first at the distinction between those creatures that engineer – we humans – and those that do not. Beavers may build dams and birds may build nests, but they do so at an instinctive level: there is no forethought or planning to these 'projects', and birds' nests today are the same as those built millennia ago, I imagine.

For our purpose of understanding the origins of engineering there have been six key steps in our evolution:

1 The first differentiation of bipedal hominids from the ancestors they shared with apes, about 5–8 million years ago. This happens in Africa.

2 The first differentiation of *Homo* from the earlier hominids, again in Africa, roughly 2 million years ago. Many species of *Homo* appear with the passage of time, some with tool-making abilities. *Homo* species disperse to many parts of the world.

3 The appearance of *Homo sapiens* about 300 000 years ago. (Some authorities consider Neanderthal man to be a member of this species, and use the designation *Homo sapiens neanderthalis*; others do not, and use the designation *Homo neanderthalis*. Neanderthal man survives until about 30 000 years ago.)

4 Within the *Homo sapiens* line, the appearance of *Homo sapiens sapiens* (that's us) about 120 000 years ago. *Homo sapiens sapiens* is found first in South Africa and the Near East. About 60 000 years ago, *Homo sapiens sapiens* builds boats and crosses to Australia.

5 The beginnings of agriculture about 10 000 years ago.

6 The start of social complexity which we know as 'civilization'.

Robert J. Wenke in his splendid book *Patterns of Prehistory: Humankind's First Three Million Years* (see ▼**Outside reading**▲) says he is

> convinced that to see oneself in relation to these transformations and the great antiquity of humankind is an indispensable part of any liberal education for students ... especially if they are [to be] engineers or theologians.
>
> (Wenke, 1999)

I agree and, although I am aware that what follows is but a trite summary of a widely researched subject, it will set our study in context (and may serve to whet your appetite).

The six stages I outlined above are not uncontroversial among anthropologists. Nevertheless, we can say that recognizably human creatures have been around for several million years. The sparse fossil record shows such creatures to have been anatomically distinct from apes, but any behavioural distinctions can only be inferred. The late Sir Grahame Clark,

▼Outside reading▲

Whilst studying the course you will occasionally come across references to writers and/or specific books that are relevant to what you are studying. These references serve to illustrate that there are often many other sources of information which delve deeper into particular topics, and which the author of a section has used in preparing the material. You may wish to pursue some of these *for your own interest* if a particular topic fascinates you. Reading references outside the supplied course material is *not* part of your study schedule, and such material will *never* be included in any assessment for the course unless you have received it from the Course Team.

sometime Professor of Archaeology at Cambridge, described human beginnings persuasively:

> When did our remote forebears cross the threshold of humanity? There is a very real sense in which man made himself – by developing his own resources he succeeded in escaping more and more from the trammels both of his external environment and from the animal appetites inherited as part of his own organism. As he added to his culture early man was confronted with a progressively wider range of choice: indeed one might say that prehistory is exciting precisely because it allows us to watch, alongside the development of material achievements, the growth of man's awareness of his own situation, an awareness that implied from the beginning an element of conscious choice.
>
> Culture in the broadest sense may be defined as a learnt mode of behaviour inherited by virtue of belonging to a social group. In this sense it is far older than the emergence of even the earliest primate: one may recall, for example, that bird song includes distinct local cultures within the general pattern of the species as a whole. Man is thus not distinguished from other animals through the possession of culture, but rather through the progressive character of his culture. The possibility of enlarging and transmitting a growing body of culture depends above all on the ability to invent and employ symbols. ... The possibility of transmitting, and so of accumulating, any considerable body of culture hardly existed, therefore, until hominids had developed words as symbols.
>
> Man's capacity for making tools to supplement his limbs is another basic distinction. Although the one presumably developed from the other, there is a world of difference between tool-using and tool-making. If we define tool-using as the use by an animal of objects from its environment to serve its own purposes we must admit that very lowly organisms could be classified as falling within this category. The great apes can use tools, and are often ingenious in manipulating sticks and strings and in stacking boxes, though their use of these is severely limited: it is not merely that they rely more upon blind improvisation than upon insight, but their activities are directed exclusively to securing visible objectives. Tool-making, on the other hand implies an element of foresight or at least a willingness to devote labour to making things ready for use at some future, undefined time. Moreover the tools themselves are not merely existing objects used for special purposes, they are artificial creations and, however elementary, they exhibit styles prevailing in the communities of their makers.
>
> (Clark, 1961, p. 35)

Among the key phases identified in this quotation is 'the growth of man's awareness of his own situation, an awareness that implied from the beginning an element of conscious choice.' To me this implies that the concept of purpose begins to have meaning for these creatures. Is purpose to be found in other animals? Birds build nests, there are ants which actually farm fungi and greenfly to provide their food and several animals use tools: chimpanzees have been observed to make them. Can you doubt that animals act purposively? Watching my cat I find it beyond doubt.

But perhaps I should think more carefully. Does a startled bird fly away from the cat for the *purpose* of avoiding being eaten? Evolutionary theory would say that birds that will fly away do so by instinct and that those which have that instinct have a better chance of surviving to breed. So the genes for that behaviour are perpetuated. Even behaviour which has been learnt, such as that of a cat stalking a bird, has its roots in instinct: the mother's instinct to teach the kitten. Only taught kittens survive (in the wild).

In order to find purpose in the behaviour of an animal, perhaps you would need, for example, to see a chimpanzee stacking boxes in order to reach food. The action has to be thought out and performed *in order* to achieve the goal. Apparently it is this ability to choose to think and reason that makes us human.

Activity 1 The first engineers

It is interesting to think about the birth and development of the human species as 'engineers'. The first audio track offers a perspective on this development. Some of the discussion which follows ties in with the audio track, but you do not necessarily have to listen to it now. You can take the audio tracks associated with this block one at a time, or listen to them together in one go. Do think about the questions which are posed by the SAQs though, even if you have to revisit them later.

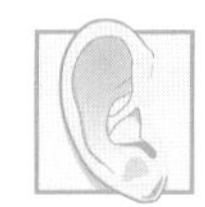

Listening to audio tracks to obtain information is a skill in itself. You should be prepared, for example, to listen to a section or sections frequently if you had difficulty understanding them. You may also find it useful to have pen and paper handy, to make notes as you listen.

You may want to review the sections of the *Sciences Good Study Guide* which cover the use of audio material in your studies: in particular, see pp. 167 and 174 of Section 3.2 in Chapter 6.

Exercise 1.1

From the audio track, how long ago does archaeological evidence indicate that humans had technological competence distinct from that of apes?

Communications, tools and cunning, products of imagination and intelligence, were the premium qualities of early humans by which they survived. Stone tools for chopping, pounding and scraping; spears and darts as weapons for hunting; and eventually the use of fire made up the technological repertoire. Those early humans lived as hunters of animals and fish, and gatherers of roots and fruits. With language they taught their children their growing knowledge of their environment and their skills for survival. Spanning the few million years of humanity, several distinct species of human have been identified, but apparently (lack of evidence prevents certainty) not making much innovation. After all, the population was so small and the available space so large that a lifestyle providing survival had no need to change.

The beginnings of an accelerating technological competence date back to some 30 000 years ago, by which time *Homo sapiens sapiens* was the only survivor of the *Homo sapiens* branch. This period of 30 000 years is less than 1 % of human existence, and the scale of engineering enterprise has taken most of that brief spell to gather pace (see ▼**About time**▲). The Pyramids in Egypt date from only 2650 BC, steam power and all it enabled is only 250 years old, and silicon integrated circuits have been with us for just a few decades. Engineering is indeed a young colt! Already it has made us a new world; perhaps already it has damaged our planet home; certainly it must carry us into the future.

This section has introduced you to what makes humans distinct from other creatures, in terms of their ability to engineer. The next section will look at how this ability has evolved and improved over time.

SAQ 1.1 (Learning outcome 1.2)

Try to think of three ways in which the *Homo sapiens sapiens* has affected the planet that are distinctly different from the ways in which any other species has affected it. Think broadly before coming up with an answer.

▼About time▲

This is the first of our asides about units of measurement. It is about time, not only because it is appropriate as we are talking history, but also because the units of time in the metric system are already totally familiar to you: though as it happens there is a story attached to that too.

Time is something that seems to happen in only one direction. We measure the passage of time for many reasons: for gauging when to eat, for knowing when to arrive at or leave work, for making appointments. Ultimately, time places a limit on our lives, and the passage of time is therefore something we are anxious to count.

For dwellers on Earth, there are three natural measures of time: the day (the time for Earth to spin once on its axis); the month (one cycle of the Moon through its phases – about 29 days); and the year (one cycle of the seasons as Earth orbits the Sun – about 365.25 days). These are naturally defined events. We could arbitrarily define a 'year' of 100 days, but the start of the year would then always be in a different season. The system of months and years that we have today has evolved over the centuries to match our counting with the natural change of the seasons. Unfortunately for inventors of calendars throughout the ages, there are no simple numerical relations between these times, so machinations such as leap years have to be tolerated.

In terms of *measuring* time as opposed to counting it, the subdivisions of the day are what are important. For measuring time we use clocks, of course, and ensuring [illegible]

defined in terms of the resonant frequency of the caesium-133 atom: specifically, the time taken for the atom to vibrate 9 192 631 770 times. (This is not a number you need to remember.)

The metric system of measurement evolved in France in the eighteenth century, being driven by the changes following the French Revolution. The Revolutionary Committee attempted to define a set of measurement standards to be followed throughout France, thereby replacing the often haphazard system of measurements of length, weight, and so on that had existed before. It gradually spread to other countries during the nineteenth century, and is now used virtually worldwide: as mentioned earlier, many of the metric measurements fall within the scope of the SI units, and it is the SI units which are recognized and used internationally.

Interestingly, when the French Revolutionary Committee was defining the metric system they decreed that the day should be divided by 100 000. This would then give a decimalized day of 10 'hours', each of 100 'minutes', each with 100 'seconds'. The story goes that French clockmakers were so geared up (literally) to making cogs for clocks working on the 24/60/60 division of the day that they steadfastly ignored the decree. The idea has never caught on since. We shall see later that this was not the first occasion that skilled French artisans had shown their power.

SAQ 1.2 (Learning outcome 1.1)

(a) How many 'seconds' would there have been in a year if the Revolutionary Committee had had its way?

(b) Calculate the number of 'real' seconds in a year.

Note that here you have to convert between different measures of time (hours and days to seconds). You should not be daunted by simple calculations such as this. Remember to use the *Sciences Good Study Guide* Maths Help section if you're unsure about any aspect of a calculation.

Exercise 1.2

Jet airliners typically cruise at 800 kilometres per hour. How many seconds do they take to travel one kilometre?

3.2 Beginning engineering: 30 000 to 10 000 years ago

Between about 30 000 and 10 000 years ago, our progenitors were still hunter–gatherers – nomadic bands following the migrations of their prey. For most of this period the climate was much colder than now, with the northern ice cap extending over much of Europe. Traces of our ancestors are to be found, for example, across the vast tract of the Sahara region, which was then a bountiful savannah grassland teeming with game. Worked flints and rock drawings of food animals can be found in places which are now desert.

From this period, evidence of ingenious artefacts survives: canoes with paddles; fishing nets, and the bow and arrow. Some knowledge and understanding of the environment of these Stone-Age people can be inferred from the ways of peoples who still live in that style – the Bushmen of the Kalahari for example. Their porcupine trap (Figures 1.2 and 1.3) is just one example from many I could have chosen.

Figure 1.2 Porcupine trap. Porcupines in southern Africa are fair sized animals about a metre long, with long bristly spines all over their backs and flanks. They follow habitual trails through scrubland when going to water. The trap is placed on a route known to be used by a porcupine

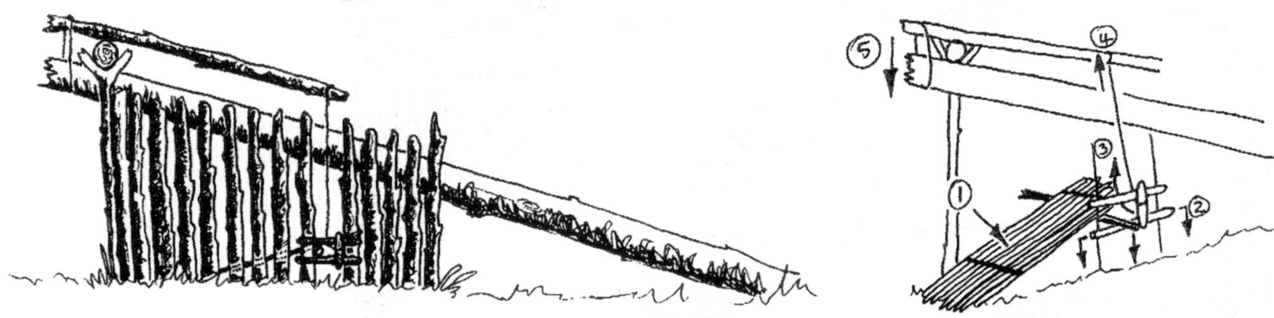

Figure 1.3 How the trap works. The porcupine enters the trap at the end where the log is raised. Inside the trap, the porcupine steps on a trigger plate of lashed sticks (1), thereby depressing a transverse stick under the trigger plate. The transverse stick bears down on the lower arm of a horizontal retaining fork at the side (2), freeing a peg (3). This untethers one end of the supporting lever at the top (4). The raised end of the log (5), now unsupported, drops onto the porcupine's back. Unable to continue forward, the porcupine tries to back up, but cannot, because its spines catch in the sticks on either side. It is stuck until the Bushman arrives

Notice how the function of the trap depends on observation of porcupine habits (where to find a porcupine, what route it follows) and on an abstract perception of what will happen when it tries to back out of the trap (the spines will stick). For my money this is engineering. It is just the sort of thing our ancestors must have invented, but of course there will be no remains of such flimsy things for us to find to confirm this assertion.

Activity 2 The beginnings of engineering

The second audio track associated with this block considers the driving forces which led our ancestors to begin using creative skills for engineering, using the bow and arrow as a particular example. You should listen to this audio track when you have an opportunity. You will be asked to pause listening whilst you think about SAQ 1.3.

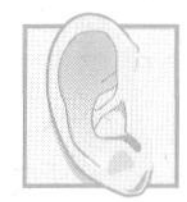

SAQ 1.3 (Learning outcome 1.3)

Imagine what factors an early people might have used in designing the first bow and arrow. Think in terms of how it must function, and how you would have to choose materials and shape them for the desired result. You are looking for answers that will guide you in making a prototype (first attempt) version of the weapon. Think of the practicalities of both *making* and *using* it.

SAQ 1.4 (Learning outcome 1.4)

The audio track has introduced you to the requirements for a successful invention: one that succeeds in the 'marketplace'. These are the contexts of *need*, of *current practice*, and of *available materials and manufacturing ability*. Which of these do you think was lacking in the case of the Sinclair C5, Figure 1.4, which prevented it from becoming a high-selling product? (The Sinclair C5 was a motorized tricycle released in about 1984. It was battery-powered, but pedals were provided for assisting on steep hills.)

Figure 1.4 Sinclair C5

At this stage we can guess that the early engineer would be pretty popular. Such a person would fulfil a useful role in the community – as a bow-maker, a trap-maker, a tool-maker, or whatever. And if everyone wants what the engineer can devise, a production system has to be organized. A labour force to cut suitable sticks for both bows and arrows, to make the strings, and to do the assembly. It might all be done in a corner of the communal cave, and hardly add up to Henry Ford style mass production, but in my short conjectural story you can see the basic elements of engineering as the provider of functional artefacts.

Figure 1.5 Development of the bow and arrow

Right up to comparatively recent times, the bow and arrow has been a favoured weapon. It continued in use even after the invention of the musket as it offered a much faster rate of fire. It has developed in many directions; Figure 1.5 shows some examples. One important discovery was that the ends of the bow should be thinned, so that it bent more uniformly and became lighter. Arrows were improved. Flint, and later metal, tips for arrows added weight and penetration without adding thickness. Adding 'flights' to arrows, made from feathers, made them fly straighter.

Designs have changed depending on where the weapon was to be used, what it was to be used for, and what materials were available to make it. For Bushmen, who learnt to stalk animals with such skill that they could get very close, a tiny bow and arrow with virulent poison on the tip was sufficient. This very light weapon was no encumbrance as the Bushmen then followed the poisoned prey until it dropped. By contrast, at Agincourt in 1415, the massed English archers firing yard-long arrows from six-foot bows of yew wood cut down the pride of the French cavalry at a range of 400 metres.

Now archery is an Olympic sport, and both bows and arrows benefit from a high degree of technological development. With scientific understanding, the most efficient profile of the bow can be modelled and calculated. Computers can be used to find the optimum shape for both bow and arrow. Progressive design such as this is one of the key features of engineering practice.

3.3 Intuitive engineering

My discussion of how the bow and arrow might have come about, as a product of ingenuity and hence as an early example of engineering, perhaps makes it seem that engineering is a simple matter accessible by anyone who takes the trouble to use their imagination. Indeed, my example of constructing shelves implies that this is so.

Nonetheless, you will learn through this course that the modern profession of engineering cannot be regarded like this: the existence of specialisms within the whole field, such as mechanical, electronic, civil engineering or whatever, implies a great deal of complexity. The knowledge and skills base of any one of those specialisms is quite enough to sustain a career. But, by going right back to the beginning I have demonstrated a starting point for engineering which in principle is indeed the sort of thing that anyone prepared to take the trouble to think and to act could do – given that first flash of inspiration. I call this type of engineering 'intuitive engineering'. You can see it in action in every shanty town around the world, or in allotment sheds if you can't go that far. Putting up shelves to your own design was a perfect example.

The intuitive engineering of the bow and arrow lies in the realization that a springy stick and a cord can be used to launch a projectile. The development of an advanced design for a bow for competition archery is something else. For one thing, the basic idea does not change: what you end up with is still a bow and arrow. What is more complicated is the level of scientific understanding applied to produce the finished article. A computer model of a bow would contain information on the material properties, the likely pulling power of an archer, and the behaviour of an arrow as it flies through the air: all quantified to allow the computer to calculate, for example, the speed of the arrow as it leaves the bow. This is clearly distinct from intuitive engineering.

Intuitive engineering has several specific qualities:

- It belongs to anyone; it is DIY engineering – the client is often the engineer.
- It has no rules or standards – so it can go wrong!
- It does not rely on scientific or mathematical modelling – though the engineer may know some of this.
- It may not even be quantitative, sizes being judged rather than measured.
- Its designs are spontaneous, developing even during the production phase.
- It may be original (e.g. the first bow and arrow) or may mimic other ideas.
- It is usually for one-off jobs.
- It uses materials which are at hand; improvise is the key word.

I will close this section with a picture, Figure 1.6. Cave paintings from this period have been found at several places. You have to imagine this work being done by the light of flickering flames deep in the cave. The vitality of the drawing connects us to these people. These paintings are creations combining artistic imagination with craft skill. The inevitable question is: Why were they made? Informed speculation suggests psychological motives – perhaps the hunters acted out scenes of the hunt to build their bravery, or at least to contain their fear; perhaps they hoped for magic to help them. We cannot know for sure, but such rituals were found in some societies when people from the developed world first met them not so long ago. For our engineering purposes the notable thing is the use of materials; again, as with the springy stick for the bow, showing selection of materials for specific function.

Figure 1.6 Cave painting from Altamira

Exercise 1.3

In SAQ 1.3 you thought about the development of the bow and arrow, an invention that was devised by an early human. However, what additional factors do you think are important for an entrepreneurial engineer to profit from an invention today? In addition to the engineer, who else may be involved in getting the product into the marketplace?

3.4 The start of agriculture

Over the past few decades a great deal of attention has been paid to the phenomenon of 'global warming', which is the slight, but quite rapid increase in average temperatures which is believed to herald major climate changes. Stasis, keeping things the same, is comfortable: we know how to cope. The uncertainty of major change is at least worrying, and may be terrifying. Even with our faith in modern technology to provide a fix for every eventuality, there is apprehension about where our world is going. And confidence in technology is not helped by the understanding that this climate change is probably caused by our own technological activities. In particular, the large scale and ever-increasing burning of fossil fuels over the past 200 years, since the Industrial Revolution, is seen as causative by increasing the carbon dioxide concentration of the atmosphere.

Why should I introduce this issue here, when I have been so concerned with the past? The answer is because global warming is not a new phenomenon on the time scale we have been looking at. About 12 000 years ago it was the trigger for the next development in human behaviour. It may not have been as sudden as modern global warming (if this turns out to be real and rapid), but it was nonetheless dramatic in its effects. The polar ice caps, which had extended almost down to latitude 50°, just south of the south coast of England, started gradually to recede around this time. Sea levels rose, rainfall patterns shifted, the temperate zones moved pole-wards. Areas such as the Middle East and the Sahara began to dry out towards what we now know as arid desert. Naturally the people followed the game to the remaining waters and became concentrated along the rivers. By modern standards 'concentrated' is a relative term, and the old way of life must have continued but within tighter geographic regions.

Activity 3 The start of agriculture and the seeds of civilization

The third audio track for the block outlines the changes that occurred in lifestyle as humans made the transition from being hunter–gatherers to being farmers. Listen to this track now (or when you next have an opportunity), and then consider the following exercise.

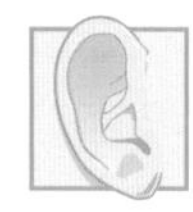

Exercise 1.4

Pause here and jot down some ideas of how life could change if you do not have to carry all that you own. What are your needs as a farmer as opposed to a hunter?

At the period I am considering, the world's human population was still extremely small and widely dispersed, so the concept of agriculture almost certainly happened independently in several places, at several times, and would only very gradually have become the dominant practice. The crescent of high land stretching from southern Turkey across to northern Iraq was one of the earliest regions for agriculture, and is important to this day as the place of origin of wheat and barley, two of our principal grain crops. In this region, the landscape is scattered with the remains of villages, the earliest dated by radio-carbon methods being in Levant and dating from 9 000 to 8 000 BC.

Sheep, goats and cattle were domesticated (dogs had probably been domesticated already) so from the earliest times farming was both arable and livestock. The same scene was set in the valleys of the great rivers (along the Nile in Africa, in the Tigris/Euphrates basin above the Persian Gulf, along the Indus in India and by the great rivers of China).

You may not think of primitive farming as engineering, and, except in the intuitive sense applied to the solution of countless small problems, such as how to harness an ox, I think you would be right not to. But the opportunity to build villages *is* engineering and begins to involve more than a single craftsman with a good idea. A community has to co-operate to make 'planning' decisions and to provide labour. There is use of a wider range of materials – timber, stone, sun-dried bricks, wattle and daub (a building material of interleaved twigs and branches plastered with mud). Particular designs emerge as favourites, presumably already reflecting the clients' preoccupation with function and the builders' with construction convenience. Although there is evidence of some trade in the rarer artefacts between settlements, each settlement would have been essentially self-sufficient. Food production was the main occupation, with the provision of artefacts undertaken in the corners of time left over. It is not yet time to think of specialists dedicated to manufacture, though doubtless there were some with greater skills whose talents were in demand.

It would be tedious, irrelevant and confusing to track through the developments little by little over the thousands of years that led from the beginnings of farming to the great bronze-age civilizations. The point is that settlement provided the opportunity for people to accumulate possessions, both for need, in pursuit of their farming lifestyle, and for comfort and pleasure. New inventions came on the scene, such as the wheel, which allowed carts to facilitate moving crops; pottery for household uses, and woven cloth made from both animal and vegetable fibres. Eventually metals were discovered (see ▼**Grasping the metal**▲). Gold was presumably the very first metal to be used, but gold is too soft to be useful, and is rare so was only prized for adornment (see ▼**Atoms, elements, and materials**▲). When copper alloys (such as bronze, an alloy of copper and tin) were discovered, possibly initially as impure forms of copper, a whole new technology of manufacture developed. Bronze products were better than flint, and so they were valued and traded. In this way, the stone age came to its end.

▼Grasping the metal▲

Of all the materials that we use for making products, metals are perhaps the most important class. They have a very wide range of properties, from soft lead that can be used as a roofing seal, to hard steels which can be used to machine softer metals. To the early engineers, metals were a godsend, being stronger than wood, and less brittle than stone.

The only problem was how to obtain them. Metals, in general, are not found lying around in handy deposits, unlike coal. This is because nearly all metals are chemically very reactive: they combine readily with other elements. You see this in the rust on your car, and the tarnish on silver and brass ornaments. Left for long enough, the iron in your car will react with oxygen in the air to form iron oxide, or rust. This is the same iron oxide from which the iron had to be extracted in the first place.

The first metal to be found was probably gold, which, exceptionally for a metal, isn't reactive (you probably know that gold jewellery doesn't rust or tarnish). Gold can be found naturally in small lumps or nuggets. However, it is a fairly soft metal, and not much use for tools. Consequently there is not much to be gained from digging it up to make a better knife. Some lucky nomads would have come across iron meteorites, but these are few and far between.

The trick of extracting metals from the ores which contain them is knowing just which rocks contain which metal. The first time that a stone fireplace bled liquid copper, which then solidified into a material that was found to be useful for spearheads and knives, was a red-letter day for our ancestors.

Other metals are much harder to extract, but the effort is often greatly repaid by the properties of the material which is finally obtained. We shall return to this topic later in the course.

▼Atoms, elements and materials▲

In order to make anything, we are constrained by the materials that are at hand. On the most fundamental level, all materials are made up of atoms, which are the basic constituents of everything around us. Atoms are tiny things: a cube of aluminium measuring 1 cm × 1 cm × 1 cm contains around 7×10^{22} identical atoms. This is a very large number. In fact, if you tried to count them at a rate of one per second, then from the answer to SAQ 1.2 we see can that it would take about 2×10^{15} years (that's 2 million billion years). Another way to think of this is that by comparison with the number of atoms in our small cube of aluminium (or any other solid), the number of seconds in a year (about 31.5×10^{6}), or even in a millennium (31.5×10^{9}), is utterly negligible. However, the number of atoms along one *edge* of the cube (about 40×10^{6}) is of the same order as the number of seconds in a year.

Although all the atoms of a lump of aluminium are identical to each other, each is different from the atoms in a lump of pure copper, say. The atoms in a piece of copper are nevertheless identical to each other. The point to grasp is that atoms come in different varieties, such as aluminium atoms, copper atoms, carbon atoms, and so on. However, the range of varieties of atom is finite, and quite small: it amounts to about 90 different sorts.

A material made entirely from atoms of one variety, for example our lump of pure aluminium, is an *element*. You will doubtless be familiar with the names of lots of elements already: iron, aluminium, gold, oxygen, nitrogen, carbon, silicon, uranium are a few. As you can see, some elements are metals, and some are non-metals.

Because the number of different types of atom is limited to about 90, the number of different elements is limited to the same number. To be precise, there are 92 naturally occurring elements, and the few extra, unstable elements that have been artificially created in nuclear reactors or 'atom-smashing' particle accelerators pushes the figure to a little over 100.

Everything that we come into contact with is made from either a single element, or a combination of two or more elements. When we say, 'That ladder is made from aluminium,' essentially we mean that there is only type of atom – the type characteristic of the element aluminium – used to make the ladder. In practice, however, particularly where metals are concerned, there are usually other elements added, maybe amounting to just a few per cent of the total, to give the best properties for the final material. Such a mixture of metal elements is called an *alloy*, and an 'aluminium' ladder will typically have small amounts of copper and zinc added to the aluminium, which greatly improve its strength. As with elements, you will certainly already know the names of the commoner alloys: brass, bronze, pewter and solder are just a selection.

Different elements have different properties, which to a degree determine the use that is made of them. For instance, copper is a better electrical conductor than almost any other metal (even gold), hence its use in cables. However, the superior corrosion resistance of gold results in its use in high-quality electrical switch contacts in applications where electrical reliability is particularly important. You will come across more on the properties of different elements, mainly as how they behave as bulk materials, throughout the course.

Exercise 1.5

The materials we use are made up of many elements: natural materials like wood are made up from carbon, oxygen, hydrogen, nitrogen, and many other elements.

See if you can identify the *elements* from the following list: tin, steel, glass, neon, helium, leather, plastic, sodium, chlorine.

We have hitherto thought of 'making things' as engineering, the bow and arrow being an example. But there is something of a conflict of terms coming up between 'craft' and 'engineering'. The various occupations of 'making things for trade' came to be conducted by specialists in particular crafts: potters, weavers, jewellers etc. In the sense that these people designed their own products, organized their raw materials and set up their production systems (including, I suppose, their tools – kilns, looms, forges), they are engineers. But as specialist artisans, once they have set up and equipped their workshop, craftspeople work in a fairly routine way, more akin to a present day machine operative than those whom we would now call engineers. That is not to decry their skills; engineers have often depended on craft workers to convert their mental creations into physical reality. This particular distinction between craft and engineering persists to our own times.

If you're unfamiliar with the powers-of-ten notation used in the input above for large numbers, see the *Sciences Good Study Guide* (the index has an entry on 'Powers of ten' which will point you in the right direction).

It seems to me that it is at this juncture that the 'profession' of engineering becomes recognizable, and is defined by the scale of enterprises that were undertaken. To build towns and cities with streets and palaces and temples needs the level of design and organizational abilities that we have identified earlier as being the role of engineers. At an intermediate scale, where shall we place the cartwrights and shipwrights who must by that time have also been specialists? You do not have to make a decision about who is or who is not an engineer. Rather, what we have perceived is that there is a spectrum of engineering running from the crudest shelf-building, intuitive engineering, through a range of skilled crafts up to the great entrepreneurial engineers such as Telford, Brunel, or Henry Ford. Talents and opportunities differ.

Through the millennia that followed the birth of civilization came a stream of innovation, such as the following:

- New materials, first copper alloys, then iron as metals for construction of machines and weapons.
- New methods of building, such as cement and the arch (both Roman inventions).
- New processes, for example metal casting, glass making, porcelain, printing.
- New power sources, for example wind and water mills.
- Mains water, under-floor heating, machines for lifting, pile driving, turning and pumping.
- Sea-worthy ships (the peopling of Australia shows that these date back a long way), culminating in the sleek Viking boats and the sloop-rigged Arab dhows.

There is an endless catalogue of achievement by exploration of what was possible.

Perhaps the biggest surprise, with hindsight, is that the principles of experimental science which were destined to open the intellectual route to the Industrial Revolution of a mere 250 years ago, took so long to come. In the sense of how engineers viewed the world they worked in, it might be argued that the 'ancient' civilizations reached right up to that time. Modernity dates from so recently.

3.5 Summary

Engineering is *imagining what, knowing how* and *getting it done*. On the Earth, at any rate, it is the sole preserve of humans, of which we are the one remaining branch, *Homo sapiens sapiens*.

The timetable of our technological evolution is *logarithmic* (see ▼Handling big numbers▲), that is to say it grows very slowly to start with and gradually accelerates. Rough dates for the significant steps are in Table 1.1.

SAQ 1.5 (Learning outcome 1.5)

(a) Find the length of time for which *Homo sapiens sapiens* has been the surviving species of *Homo sapiens*. For what fraction of this time have we been 'civilized' in terms of city dwelling?

(b) For what fraction of the time since we became 'civilized' (city dwelling) have we been 'industrialized'?

(c) Can you imagine living without electric power? For what percentage of the time since the appearance of *Homo* has this boon been available?

▼Handling big numbers▲

If you look at the numbers in Table 1.1, you'll see that they go from 8 million years ago to about 20 years ago. The jumps in between are quite uneven, from 60 years to 3–5 million years.

Variations like this can't be expressed easily using a linear scale, which is one that goes 10, 20, 30, 40, … Using increments, like those, it would take a long time to reach 4 000 000! On the other hand, if we use large

Table 1.1 When and what for humanity

When	*What*
5 to 8 million years ago	Bipedal hominids appear.
2 million years ago	*Homo* genus appears
300 000 years ago	*Homo sapiens* species appear.
120 000 years ago	*H. sapiens sapiens* appears.
30 000 years ago	*H. sapiens sapiens* is the only surviving member of the *Homo* lineage.
10 000 years ago	Global warming (preceded by 5000 years of fluctuating temperatures). Start of deliberate seed sowing. Self sufficient farming villages widely spread – little interaction. 'Intuitive' engineering for housing and agricultural tools.
5000 years ago	Cities with dependent hinterland flourishing. 'Professional' engineering necessary for major civil works. Specialist 'craft' production of artefacts. Metal working beginning. Trade can provide livelihood. Political hierarchies govern large populations.
3000 years ago	Iron working starting.
2000 years ago	Roman Empire controls Europe and Mediterranean. Military motivations for engineering works (roads).
700 years ago	Printing invented in Europe – an IT revolution!
200 years ago	Effective steam power available – industrialization of production in full swing. Major civil engineering for transport.
80 years ago	Electric power and internal combustion engines widely available. Telecommunications (radio).
1980s	Computers widely available; networking beginning.

increments, such as a million or 500 000 years, we lose detail, because events that were hundreds or a few thousands of years apart start to look as though they were simultaneous. As Figure 1.7 shows, the last 10 000 years of human history cannot be sensibly plotted on such a scale if it also includes the appearance of humans 3–4 million years ago.

A better solution is a scale where equal steps correspond to leaps of ten *times*: 10, 100, 1000, 10 000, 100 000, 1 000 000 (Figure 1.8). Such a scale is known as *logarithmic*. This can give a clearer picture of what is happening, and allows details at both the small numbers and the big numbers to be distinguished. You will come across this way of expressing large spreads of numbers fairly often.

More information on powers of ten can be found in the *Sciences Good Study Guide.*

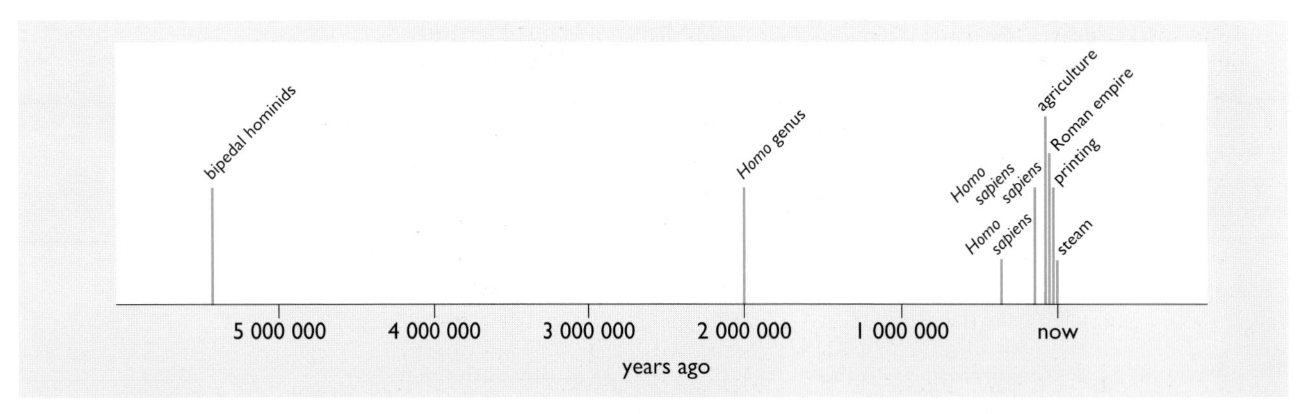

Figure 1.7 A linear scale

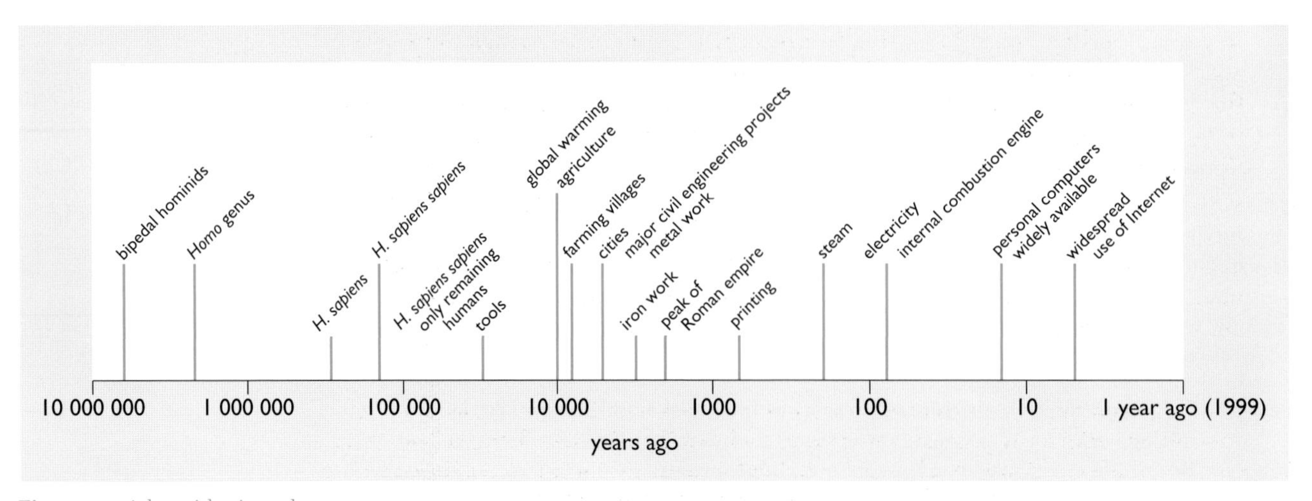

Figure 1.8 A logarithmic scale

4 What is engineering?

4.1 Looking for a definition

At six o'clock this morning my manufactured alarm radio awoke me in my manufactured bed. I went to my bathroom, with its manufactured fittings, washed with manufactured soap, dressed in my manufactured clothes and came down the constructed stairs to eat my manufactured breakfast. My selectively bred (but otherwise unmanufactured) cat greeted me, and I opened a manufactured can of manufactured food for her; she seems to find domestication congenial. Outside, in the old tree in my garden (all organized, hybridized and fertilized by human intervention) a cock blackbird called his territory: the first wild thing of the day.

Look around yourself. Recognize just how much the material environment we live in is of our own making; it has been *engineered*. Even to provide the simplest of my examples, the soap, has required, if you think about it, interactions among lots of people engaged in an enormous range of activities.

Exercise 1.6

Soap is made by reacting a fat (often a vegetable oil) with an alkali (a generic name for a type of chemical).

In three minutes, write down as many things/people/activities as you can think of that are involved in getting a bar of soap from raw materials to the block in your bathroom.

These simple examples are sufficient to remind you of the great breadth of 'economic activity' which supports our modern industrialized lifestyle. Perhaps because we have been born into this 'civilization', rather than chosen it, the depth of our dependence on its intricate, interlocking systems of supply is not so obvious. We expect our food to appear in shops, water to be on tap, sewage to disappear without thought or effort, power to come from a socket in the wall, communication to need but the touch of a few buttons.[1] We expect to be provided with shelter from the elements, to be entertained, transported, kept in health and defended. For our labours in some capacity within these systems, society rewards us by allowing us to participate in the providence. (The hungry, the homeless, the jobless, the victims of crime may point out that the systems are not open to everyone or not always successful in their action. There are politicians, moralists and the greedy who would say that that is quite as it should be too.)

It is engineering that puts all this in place. All the material things from bars of soap to space shuttles that are the apparatus of those provident systems are the products of engineering. People have recognized a need for some function to be achieved (e.g. de-greased skin), thought out or discovered how it can be done (soap), designed, set up and run the means for making the thing (raw materials, process equipment, energy source, labour, finance) and delivered it to the consumer. You can see that this nutshell description of what engineering involves already identifies or implies quite complex interactions; many different people could be engaged in different bits of the scheme. There are so many things to be decided, such as the following.

[1] The 'millennium bug' threatened to disrupt this system of life-support, as computers controlling various key processes were expected to fail as the date changed from 1999 to 2000. In the event, not much happened, but for many people our dependence on the various systems of supply and support was strongly emphasized for the first time.

- How is the function 'need' identified?
- Who invents solutions?
- How is the decision made to accept a particular solution?
- How is it decided to set up a production facility?
- Who designs that production facility; what production level; what resources?
- Who builds it, runs it, pays for it?
- How is the product costed; will it succeed in the market?

Different societies at different times have different answers to these questions. In our capitalist society you'll find a host of professions getting in on the act: market researchers, bankers, lawyers, steel erectors, brick layers and joiners, machinery sales people ... and engineers. So who or what are engineers; what do they do?

Ask this question of a dozen people in the street and you are likely to get a dozen answers. Some may think of building things, such as bridges or skyscrapers; some may think of repairing things like cars or televisions. Others may relate the word to common phrases or job titles, such as *genetic engineering, software engineering* or *civil engineering.* Yet another may tell you that engineers run factories.

If you ask 'What is engineering?' of a dozen professional engineers you'll also get many different answers. Their responses will depend on their particular backgrounds, experiences and jobs of the moment. Some will say that engineering is about making things, or developing things. Others will answer in more general terms, saying that it is about generating profit for a company or wealth for a country, or about improving the quality of life. All these answers would be correct; but every simple definition of engineering leaves something out.

Clearly engineers are responsible in several ways for making some of those decisions we listed for the would-be soap-maker. Engineers handle the material aspects of the business. They will know (or be able to work out) how to design a factory suited to making soap, what plant and which raw materials to buy, how to work out what power supplies will be needed, and so on. They may be able to calculate how much all this will cost, but they probably will not control the decision of whether to invest that much in the venture. Nor will they be asked to plan the advertising campaign.

You might like to look for a definition of engineering in a dictionary or encyclopaedia and see whether or not it agrees with your own perception of what it is. By the end of the course you will hopefully have a clearer understanding of the field which will enable you to give a comprehensive answer; and it will take you several sentences to describe, I expect. But for now we need to take one of those simple, incomplete answers in order to progress. Let us try a very general definition of the engineer's task. I have adapted the following from several published definitions to reflect my perception of what an engineer does:

> To apply scientific and technical knowledge to the design, creation and use of structures and functional artefacts.

There is a lot wrapped up in this definition. One of its key words is *design*: there is no point in amassing scientific knowledge if it cannot be applied to develop a new idea to fruition, or be used to improve an existing product. Design is a key underpinning to all of engineering.

The following section explores this definition through some real examples.

4.2 Some case studies

Which examples from engineering practice will best help us illustrate our definition of engineering? This definition can carry us quite a distance in seeing what engineers *do*. Some examples will illustrate the point. But which to choose? Anything out of our whole history of bending the environment to fit our whims will do. For a fuller discussion I have picked four items: the Pont du Gard, a beautiful Roman bridge in the south of France; a disposable ballpoint pen; muskets; and the plant for making the chemical ammonia. This set will enlighten us concerning varied aspects of engineers' practice for, apart from their intrinsic interest, the case studies in this set demonstrate the distinction between one-off engineering solutions and mass-produced solutions.

4.2.1 The Pont du Gard: one of a kind

Crossing ditches, dips in the land, streams, rivers and roads is one obvious engineering task. Most of us will cross several bridges during a normal day, and there's sure to be one not far from where you live. You will be aware that there are many different designs of bridge, from little more than a beam across a gap to elegant suspension bridges.

For the makers of early bridges, such as the Pont du Gard (Figure 1.9), the problem was that they were limited to materials like wood and stone. Metals, although in use for tools, armour and weapons, couldn't be produced in sufficient quantity or quality for bridge building until the nineteenth century.

Wood doesn't last too well, so that brings us to stone. One of the problems with stone is that it is *brittle*: it is easily broken by an impact, and will tend to break rather than just deform. Stone is an example of a ceramic material. The pottery mugs you have at home are also ceramics, and they break easily if dropped onto a hard surface; a metal saucepan, on the other hand, would not break, though it might end up with a dent – metals tend to be *tough*.

Figure 1.9 The Pont du Gard

When using stone, the trick is to ensure that it is used so that it is being compressed (see ▼Compression and tension▲). Think about the bricks used to make your house: they are stacked one on the other, so that each brick is compressed by those above it. This works fine. However, building a bridge is different from building a wall. A way is required to minimize the areas which are in tension. The solution in the case of the Pont du Gard is to use arches: curved columns of stone which are compressed by the span of the bridge above them.

Unlike most bridges, the Pont du Gard was not designed to carry people or animals, although tourists with a good head for heights can walk across it. It was built under the patronage of the Roman Emperor Agrippa in 18 BC to carry water to Nîmes in southern France. The water channel is nearly 50 m above the river, and traverses a distance of 270 m. The channel is nearly 2 m^2 in cross-section and was lined with cement to make it waterproof. The arches, however, were all constructed of accurately cut stones using no cement at all.

▼Compression and tension▲

In construction, particularly, and many other areas of engineering, the forces acting on the materials are critical to whether or not a structure will be safe. Materials have limits of strength which must not be exceeded.

Forces are either compressive or tensile. You can think of compression as a 'squeezing' force, and tension as a 'pulling' force (see Figure 1.10). Modern structural materials (mainly metals) can withstand both tensile and compressive forces. Stone and similar ceramics are fine in compression, but their strength in tension is much lower.

The simplest sort of bridge would be a slab over a ditch (Figure 1.11). When there is something on the bridge, it will bend: even if only very slightly. Bending puts tension onto the bottom of the slab, and compresses the top (Figure 1.12). So in building stone bridges, this tensile force must be minimized below the level where the slab would crack. This can be done by making the slab shorter between its supports, or thicker. Look how the design for the Pont du Gard has plenty of supports along its length, so that none of the stones will bend unduly. Bending crops up a lot in engineering: every time you walk across a room you bend the floorboards beneath you. You will come across this again in more detail later in the course.

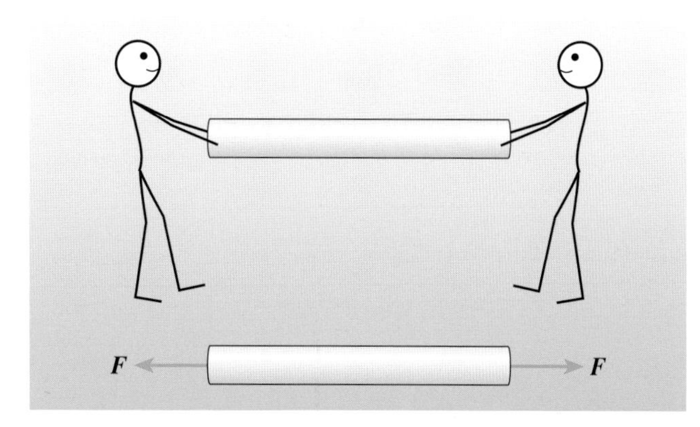

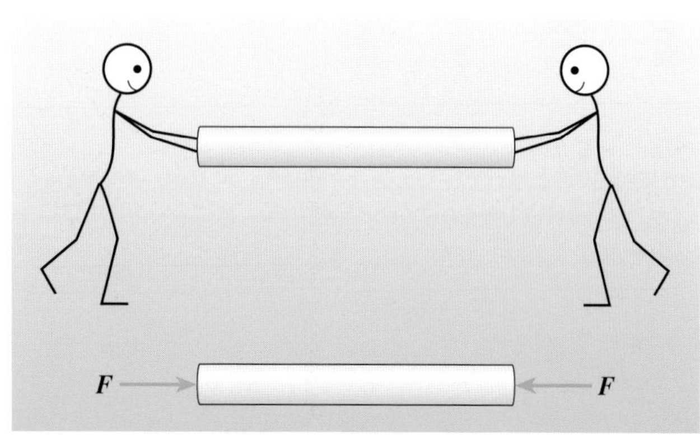

Figure 1.10 Tension and compression

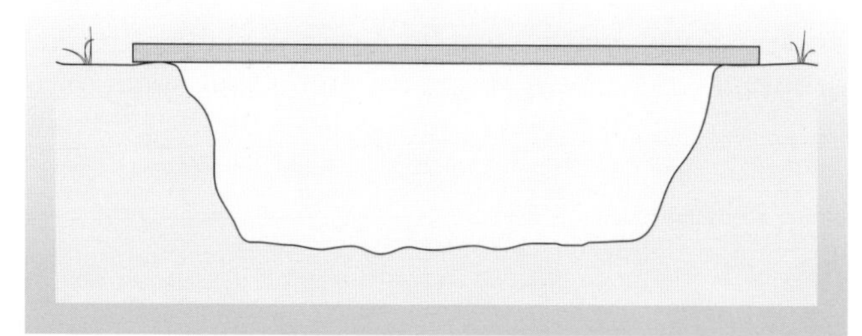

Figure 1.11 Slab over a ditch

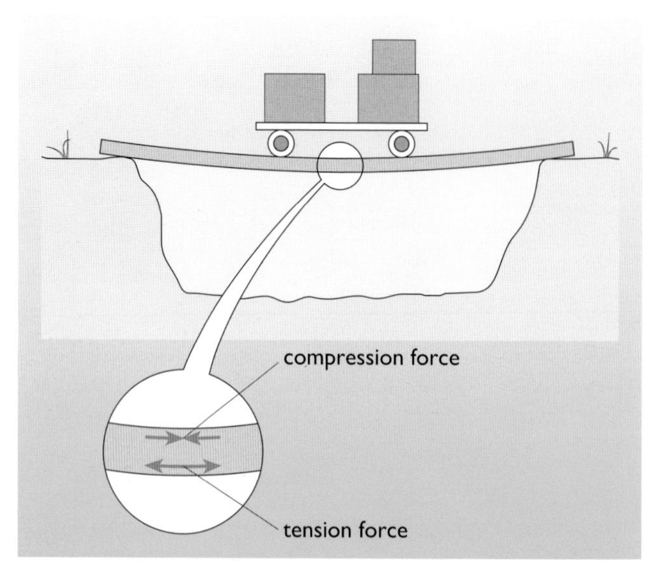

Figure 1.12 A cart on the slab causes the slab to bend. The forces in the slab are shown in the detail: compression at the top and tension at the bottom

With the Emperor backing the construction of the bridge, the financial side of the enterprise was presumably not a concern. But consider the necessary organization of materials and labour. First it had to be decided exactly what was to be built. The design had to be worked out in meticulous detail so that the water channel would be at the correct height with the correct fall, so that the bridge would fit the site, and so that the stones could be cut to fit to form the arches. The very existence of the bridge is proof that the Roman engineers had accurate methods of measuring and could transfer their calculations in instructions to the artisans: see ▼**Measuring sizes – length**▲. A quarry had to be established and means of bringing stone to the site provided – appropriate roads and carts. Workshops for preparing the stones to prescribed measures would most likely be on site so that flexibility could be maintained during construction. Then there was a need for plenty of timber and skilled carpenters. Arches are built on a 'centring', which is a timber scaffold in the

▼Measuring sizes – length▲

The metric system was defined in France in the 1790s, following the French Revolution, to bring order to a confusion of vague and inaccurate standards. The old units were well defined – officially – but petty corruption and fraudulent trade made a new standard necessary.

Units of length

The SI unit of length is the metre. The original intention was for it to be simply related to the size of the Earth. The distance from the north pole to the equator along the meridian through Paris was to be ten million metres. This distance could be measured by astronomical methods [illegible] he [illegible] bar, [illegible] ry [illegible] etre [illegible] pies

[illegible] ing [illegible] ss you [illegible] s a [illegible] d a

[illegible] r units [illegible] ed or [illegible] or [illegible] 3.

[illegible]

*deca*metre	×10	
*hecta*metre	×100	
*kilo*metre	×1000	km
*mega*metre	×1 000 000	Mm
*giga*metre	×1 000 000 000	Gm

Table 1.3 Fractions of the metre

	Factor	*Symbol*
*deci*metre	×1/10	dm
*centi*metre	×1/100	cm
*milli*metre	×1/1 000	mm
*micro*metre	×1/1 000 000	µm
*nano*metre	×1/1 000 000 000	nm

The prefixes shown in *italic* in the left column are common to all metric units to describe the multiplying factor to be applied to the base unit. By agreement the *Système International* of units (SI) recognizes factors going up in steps of one thousand. Thus nm, µm, mm, m and km are SI units. Indeed, apart from the obviously useful centimetre (bigger than a third of an inch, but less than a half) the others are not in common use in everyday life in the UK at the time of writing. The km is the standard measure for distance in other European countries; other units such as the mm, µm and nm find extensive use in engineering measurement: as you will see.

To get some idea of scale, kilometres are useful for measuring place-to-place distances (e.g. London to Paris is 340 km). By definition of the metre it is 40 Mm around the Earth. It's about 400 Mm to the Moon, and 140 Gm to the Sun. On the small side, a millimetre is about half the height of this letter x. A micrometre (also colloquially called a micron) is almost as small as can be seen with a good optical microscope, and a nanometre is taking us towards the size of atoms. You may have heard of 'nanometre or nano-scale engineering'. It is to do with layered electronic devices where layers just a few atoms thick are deposited.

SAQ 1.6 (Learning outcome 1.1)

(a) Which is the larger distance in each of the following pairs?
  (i) 2000 mm or 1.8 m
  (ii) 1 mm or 1 km
  (iii) 100 µm or 0.1 mm.

(b) There are 25.4 mm in an inch (Imperial measure). How many inches are there in a metre?

(c) In a photograph taken using a camera attachment with a particular microscope, features that are really 2 µm wide appear as 2 cm wide. What is the magnification?

More information on units and practice of converting between different powers of tens can be found in the *Sciences Good Study Guide*, Chapter 5, Section 4.

shape of the arch. Only when all the stones are in place for the whole curve of the arch can the centring be removed; the stones drop slightly and the arch becomes stable under its own weight. You will also realize that some heavy weights, both stones and timber, have to be shifted – which calls for ropes, pulleys, levers .... And day by day the project had to be supervised, or kept on track, by an engineer who understood what to do and how to do it. Evidently this engineer did it rather well; the result is still standing over 2000 years later.

4.2.2 Disposable pens and mass production

The idea of writing – setting language into a visible and permanent form through a set of symbols – is young relative to the age of our species, only a few thousand years. Among its earliest forms was the cuneiform script. The writing was a series of wedge-shaped (that's what cuneiform means) indentations in tablets of damp clay. When fired, the clay became solid, so making a permanent record (Figure 1.13). Tonnes of these have been excavated by archaeologists and deciphered to reveal the records of the bureaucracy of the Sumerian and other Mesopotamian civilizations.

Figure 1.13 An example of cuneiform writing

The Egyptians invented paper – the word derives from 'papyrus', the reed of the Nile which they used for paper making – so inks were needed in order to make a mark on the paper, with a 'pen' to carry the ink onto the paper in just the right quantity for legibility. In China the pen was a bamboo stick, shredded at one end to make a brush. Thanks to the skill of some brush users, calligraphy became an art form, one of the four Zen skills. Another simple pen was the quill, just a large feather cut across the stem at an angle and split to form a nib.

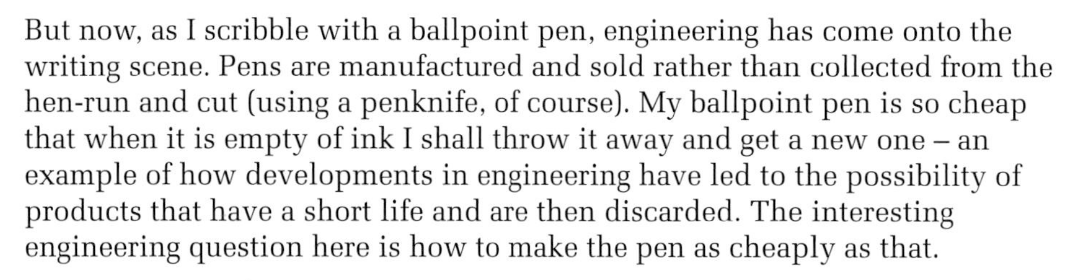

But now, as I scribble with a ballpoint pen, engineering has come onto the writing scene. Pens are manufactured and sold rather than collected from the hen-run and cut (using a penknife, of course). My ballpoint pen is so cheap that when it is empty of ink I shall throw it away and get a new one – an example of how developments in engineering have led to the possibility of products that have a short life and are then discarded. The interesting engineering question here is how to make the pen as cheaply as that.

Again, it starts with design. Obviously the thing must be designed to function as a pen, but the cost of its materials and its manufacturing process must be carefully thought about in order to get its cost very low. How is it done?

The lower part of Figure 1.14 is my pen; the upper part is a longitudinal section (see ▼**Engineering drawings**▲). It is not a unique design, and you will know that there are many similar designs in use. Indeed, it is one of the interesting features of design that many solutions can be conceived within the boundaries of a specification – a theme which will be taken up later in the course. What all such designs have in common is the use of a ball to transfer ink to the paper as the ball rolls across the page.

The pen in Figure 1.14 is made of six solid parts plus the ink. The outer barrel and a lid which fits over it are made of two different types of plastic. The metal piece contains the ball, and both the outer barrel and the ink tube fit its other end. The plastic end-plug prevents ink loss from the back of the pen and holds the far end of the ink tube. Each part has to be made separately and then they have to be assembled.

▼Engineering drawings▲

Figure 1.14 gives an example of a cutaway drawing: a representation of what you would see if, in this case, you were to slice the pen through the middle. It's easier to illustrate the different parts on a diagram like this than it would using a photograph, or just words. It also allows us to show clearly the dimensions of the various parts, by marking these on the diagram. You should not find diagrams like this difficult to decipher: you will come across others during the course.

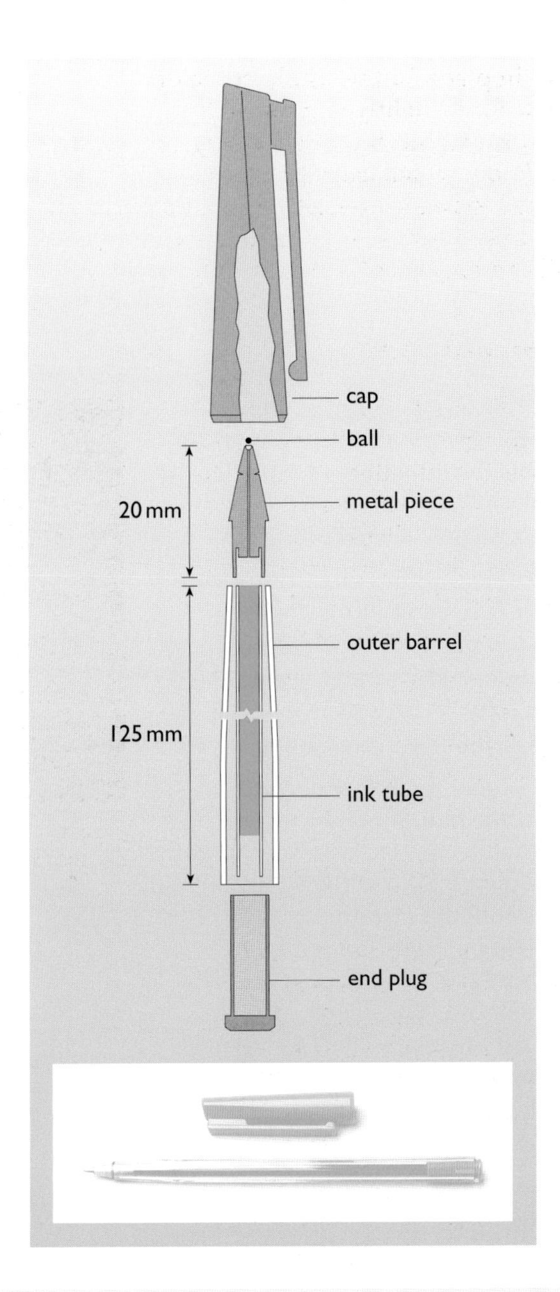

Figure 1.14
Ballpoint pen

SAQ 1.7 (Learning outcome 1.6)

What are the problems that must be overcome in order for a ballpoint pen such as the one in Figure 1.14 to work successfully? Put yourself in the position of an engineer who has been given the basics of the design: ball, ink, and barrel, and has to solve the problems enabling it to be made. See if you can come up with three problems.

The engineering secret that enables such a thing to sell in the shops for 20p is to devise machinery that will make millions of pens. A million pounds invested in machinery that will make fifty million pens gives a unit cost of 2p per pen. That leaves 18p for materials, distribution costs and profit. My sums are probably in the right ballpark, but much more complete and accurate costings would be done in reality. For instance, *exactly* how long a period shall the tooling costs be spread over? What about buildings and labour costs? And how does the VATman come into the equation?

With the advent of low-cost plastic materials provided by the chemical industry, using oil as feedstock, such cheap, disposable objects have become open to manufacture (see ▼**Plastics and polymers**▲). Thermoplastics, which are a type of polymer which soften when heated, can be moulded rapidly and relatively easily to close dimensional tolerances. The outer barrel, cap and

▼Plastics and polymers▲

One of the material revolutions of the twentieth century was the advent of polymers, which are artificial materials made from various by-products of the oil industry. The more common name for these materials is 'plastics'. The words 'plastics' or 'polymers' describe a range of materials in many different applications: plastic drinks bottles, plastic cutlery, plastic toys, plastic boxes.

The word 'plastic', however, predates the discovery of polymers by a few centuries, and originally referred to the capacity of certain materials, such as clay or wax, to be moulded or shaped. 'Plastic', in this original sense, describes the properties of some, but not all, polymer materials. Nevertheless, the word has passed into common usage to refer to almost any polymer.

end plug will be so made. The ink tube is also plastic, but being of constant cross-section, can be made even more rapidly by extrusion (that is, squeezing through a suitably shaped gap – think of toothpaste coming out of a tube). The tube can then be filled with ink and cut to length. Both these processes involve molten plastic being forced into a die which produces the required shape.

The writing end is more tricky to make. A typical ballpoint pen will inscribe many kilometres of writing during its use, hence the material for the ball needs to be hard, so that it does not wear away too quickly. Fortunately, tungsten carbide (a hard ceramic material) is often used for ball-bearings, and these can be made to a very close tolerance. Thus the balls for the pen can be bought ready-made from a ball-bearing manufacturer.

The pen maker has to make the metal piece to receive the ball with just enough clearance to allow the ink to flow round the ball. It can be turned on a lathe from bar stock. If this were to be done by hand, the pen would become expensive simply because of the labour cost involved. So it is likely that a computer-controlled lathe would be used. To enable the machining to be done at high speed, for rapid output, a material that can be easily cut, and that is not brittle (so it won't chip), must be chosen. Brass (an alloy of copper and zinc) meets the bill, but is quite an expensive material. The design/costing exercise thus has to balance material costs against tooling costs. The balance will depend upon the quantity of brass used and the time needed to machine each piece. If you look at the metal piece which holds the ball at the end of a cheap ballpoint pen, you will see that the end is indeed a yellowish metal: brass.

A production sequence is needed which will provide the top metal part, insert the ball and swage the end over to trap the ball. Finally, all the parts have to be assembled. Once again, a machine must do this in order to maintain a high production rate. So a critical factor is that *all parts of each sort must be interchangeable.* It would be useless, for example, if a batch of end plugs came to the assembly machine with the centring ring too big to fit into the outer barrel. So this brings us on to the key part of this case study: mass production.

4.2.3 Muskets and mass production

Achieving interchangeability was an essential breakthrough to mass production. Look around your room right now; you will see many artefacts which have been assembled from parts: all your furniture for example (unless your tastes are particularly *avant garde*). Whether the parts have to be made to fit accurately, as with the ball in the ballpoint pen, or can be a sloppy fit with gaps filled with glue (as with furniture, if gaps and glue are hidden from the eye) will depend on the function to be met by the product.

Close-tolerance interchangeability was achieved in manufacturing only after a long and stony path. It is now a characteristic difference between industrial and craft manufacture; essentially it can only realistically be achieved by machine-made parts. Tremendous efforts were made towards interchangeability in the state armaments industry of eighteenth-century France, but with very limited success. The infantry weapon of the time was

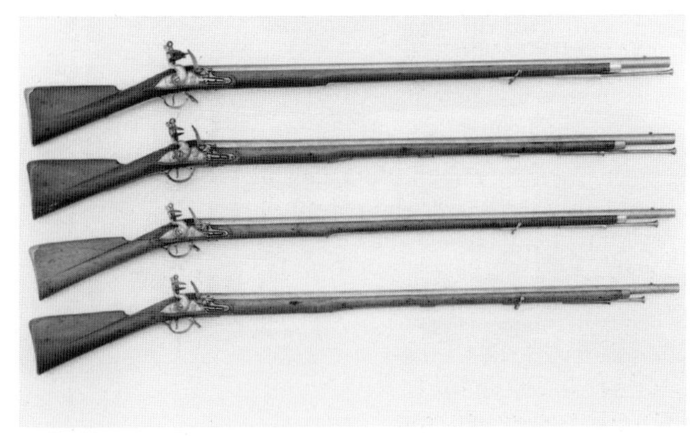

Figure 1.15 English musket from the same era as the French M1777

the muzzle-loaded musket, where the gun was loaded by pouring gunpowder down the barrel, and shoving the ball (the precursor of the modern bullet) down on top. The gunpowder charge was detonated by a spark caused by a flint striking a steel plate, and the 'flint lock' was a critical sub-system of the gun.

The lock was a complex assembly of levers, pivots, plates and springs – some twenty parts in all. To equip a large army, many thousands had to be made, and this task fell to the esteemed profession of the locksmith. (Note that the 'lock' in 'locksmith' refers to the gun mechanism, rather than the door-fastening mechanism of modern usage.)

Master locksmiths first forged the parts from iron stock, and then complete locks would be made by sorting through piles of component parts to find a combination that could be made to fit together and work – with some judicious application of the file along the way. Thus, to make a working lock required a good deal of hand-filing and assembly (using hired labour); each lock was an individual mechanism, and the parts were rarely interchangeable between locks. Imagine how unfeasible such methods would be for the manufacture of the ballpoint pen.

Production rates by such methods were very slow, so there were several hundred craftsmen in the trade working in small groups in workshops scattered around the armament factories. Locksmiths were highly skilled, and would protect their own interests, which they naturally saw as preserving the status quo. However, in 1723,

> at the Hôtel des Invalides, in front of august witnesses, Guillaume Deschamps disassembled fifty flintlocks and recombined their parts to produce fifty functioning flintlocks. The minister of war ordered Vallière to supervise the inventor's attempts to expand his system. By 1727 Deschamps had manufactured 660 locks judged interchangeable by Vallière's own inspectors, 'all properly reassembled without a single stroke of the file'. Each lock, however, had cost five times the current price. Undaunted, Deschamps proposed a larger manufacture to produce 5000 identical gunlocks, which he expected to cost only one twentieth of the current price. Tasks were to be divided among specialist workers, who would use dies, gauges and filing jigs to shape parts precisely. Master pattern locks would be distributed to the War Office, Grand Master, inspector, controller and examiner. Deschamps noted that with six hundred locksmiths in the St. Etienne region, he could locally staff a manufacture to produce 40 000 gunlocks a year.
>
> (Alder, 1999, p.221)

In fact none of this came to pass. In spite of the advantages of the new process, both in the initial assembly of guns and in their subsequent repair, especially in the field, it was by no means clear that productivity could be increased or unit costs brought down. Indeed it took locksmiths *longer* to make parts which all matched the gauges than to make sets which could be individually assembled functionally. With large quantities of parts being made by the existing routes, it was always possible to find a set of components that would fit together suitably, even if this meant no interchangeability from one lock to the next. The locksmiths were also less than co-operative towards the introduction of the new process, fearing that their craft was being de-skilled so that they would lose their powerful position. Nor were the administrative and technical problems of organizing the dispersed labour force easily solved.

Fifty years later we read of Honoré Blanc, controller of musket production at the three factories in France, repeating the demonstration of interchangeability with the locks of the standard French musket known as the M1777. The same arguments raged and little was done. Eventually, in post-revolutionary France at Blanc's own manufacturing base at Roanne, interchangeability of flint-lock parts was achieved on a large scale. New organizations of the processes, new divisions of labour and new machines contrived to keep the costs of interchangeable locks to no worse than 30% greater than the individual locks made by the old methods. When Napoleon's armies marched across Europe, the bases of mass production had been set.

4.2.4 Ammonia synthesis by bulk production

To make high explosives, such as TNT, requires chemicals called 'nitrates'. At the time of the first World War these were obtained from a natural mineral, which was mostly supplied from South America. Germany was cut off from these mineral supplies, so its chemical industry was charged with the task of manufacturing nitrates artificially.

Nitrates are essentially composed of two gaseous elements: nitrogen and oxygen, which are chemically combined. Nitrogen is abundant in the atmosphere, around 80% by weight, but in an unreactive form. The task therefore came down to coaxing atmospheric nitrogen into a more reactive condition. A particularly good method of achieving this is to combine nitrogen with another gaseous element, hydrogen, to form ammonia. The Haber–Bosch process for making ammonia was the outcome of the German war-driven research and enabled Germany to maintain its supplies of munitions. The process is now the mainstay of ammonia production world-wide, with ammonia used almost entirely as the basis of nitrogenous fertilizer.

The production of a chemical like ammonia is rather different from the mass production used for a ballpoint pen. This is because the product, ammonia, is not particularly useful in itself: it becomes the starting material in yet further processes. The product does not come out as discrete items, and production may be continuous.

Within the context of this study I want to use this as an example of engineering as the 'appliance of science', the science in this case being chemistry. (See ▼**Engineering with atoms**▲.)

For the production of ammonia, the elements nitrogen and hydrogen have to be coaxed to react together: ammonia is the product of this reaction under the right conditions. The right proportions are needed also: a tonne of ammonia requires 820 kg of nitrogen to react with 180 kg of hydrogen (see ▼**Units of**

▼Engineering with atoms▲

When chemistry is applied as a means to an end, where that end is a product with a function, it can be thought of as engineering with atoms. Clearly it's large-scale engineering, not one atom at a time, and the chemical reactions involved often progress without any intervention from humans: indeed, some reactions can only be prevented by ensuring that the chemicals involved are kept well apart. However, there is still engineering skill required for chemical manufacture, in many cases, to create the right conditions for bringing atoms together to produce the required product, be that a drug, a fertilizer, or furniture polish. There is design of the production plant to consider, in particular.

True 'atomic engineering' is still some way in the future, but we are already progressing towards this goal. Figure 16 shows a cage of cobalt atoms which may form the basis of atom-scale electronic devices of the future.

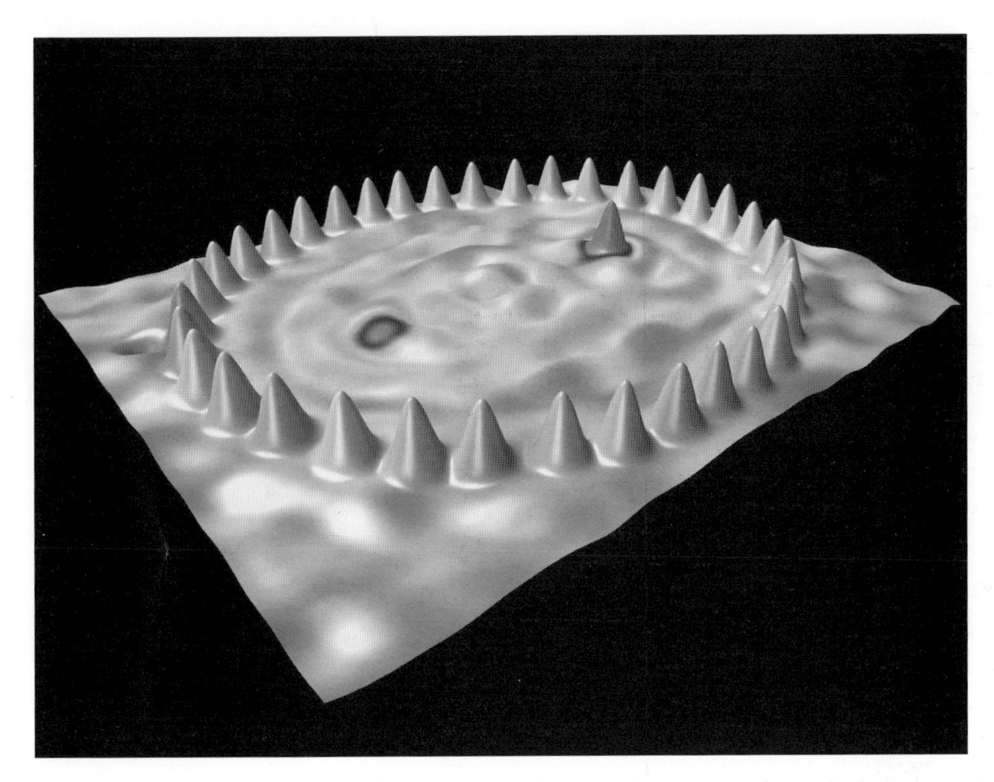

Figure 1.16 'Quantum cage' of cobalt on copper. Each cone in the ring is a single cobalt atom

mass▲). The reactants (i.e. the chemicals that are reacting) are gases, and so is the product. So one constraint on the production system is that it has to supply the ingredients and remove the product in the gaseous state.

Theoretical chemistry provides us with a clear understanding of what conditions are needed to produce ammonia. Fortunately, the principles are simple to understand. It is worth looking at their consequences here as an example of the fundamental physical laws that constrain much of engineering. Making ammonia is not as simple as stirring the raw materials together in a bucket, and understanding the process by which ammonia is formed allows us to make it as efficiently as possible.

First, the reaction between nitrogen and hydrogen proceeds better at high pressures. If the engineer can provide a high-pressure system for making the ammonia, it will be much more efficient, and so more productive, than one operating at normal atmospheric pressure.

Secondly, chemical reactions often produce heat, and the formation of ammonia certainly does this. However, if the temperature becomes too high, the reaction can become slow; so the engineer can help production by getting rid of the heat produced by the reaction and keeping production high.

Thirdly, it's important to extract the ammonia as it's being made. As more and more ammonia is produced in the reaction chamber, using up the nitrogen and hydrogen, the reaction will slow down and eventually stop. This would happen before all the nitrogen and hydrogen were used up: the presence of the ammonia slows the reaction. Extracting the ammonia product and pumping in more ingredients will keep production continuous.

This is an example where the engineering is driven by the chemistry. Without an understanding of *how* ammonia is made by this reaction, and how chemical reactions behave in practice, it would be an extremely inefficient process, and indeed perhaps impossible to manufacture ammonia on any viable industrial scale. Yet the chemical knowledge is insufficient without the engineering expertise to build plant to the required specification: plant which will accomplish those things which I have outlined above.

▼Units of mass▲

The basic unit of mass is the **gram**, but it is actually the kilogram (1000 grams) that is standardized. A litre of pure water was *defined* to have a *mass* of 1 kilogram (kg). Again the definition was translated into a single prototype, this time as a cylinder of platinum, which is a dense, precious and uncorrodable metal. Water itself is no good as a defined standard because it can evaporate (and become lighter) or dissolve substances from the air (and become heavier). Since volume-for-volume platinum is about 20 times as heavy as water, it works out that the standard kilogram is actually quite a small cylinder; about 4 cm high and wide, housed in Paris.

The chosen definition was hideously difficult to translate accurately into the metal standard. First, water is difficult to get really pure – it dissolves almost anything, including atmospheric gases, to some extent. Then there is the problem of making a container which holds exactly a litre. Both the container and the water will change volume with temperature, so that has to be set. Finally a truly accurate balance had to be made. Jobs for engineers here! Not surprisingly the standard kilogram *is* the mass of that jealously guarded piece of platinum in Paris. A litre of water at 20 °C weighs the same. (Note that the temperature needs to be defined also; materials change their volume slightly with changes in temperature, and a litre of water measured accurately at 20 °C will be slightly more than a litre at 21 °C and slightly less at 19 °C.)

A bag of sugar typically weighs a kilogram, or half a kilogram if it's a small bag. Once again, as with length, multiples and subdivisions of the basic unit are useful; and, because the SI consistently uses a factor of 1000 to define multiples and subdivisions, the prefixes that we met in connection with length are also used with mass. Table 1.4 sets them out.

Notice that 1000 kg (a megagram) is actually called a tonne. (Its spelling distinguishes it from the Imperial 'ton', though they are very close – to within about 2%.)

You will find examples of these in everyday situations. Listen out for figures describing air pollution in micrograms per cubic metre (μg m^{-3}). The legal limit for the concentration of alcohol in a driver's blood is 80 mg per 100 cm^3. Chocolate bars will be sold as weighing so many grams.

Table 1.4 SI units of mass

Unit	*Quantity*	*Symbol*
microgram	10^{-6} g	μg
milligram	10^{-3} g	mg
gram	1 g	g
kilogram	10^{3} g	kg
tonne	10^{6} g	t

Mass and weight

I should finish this section with a few words about the distinction between 'mass' and 'weight'. The confusion of mass and weight arises because we say that the block of platinum *weighs* a kilogram, conjuring images of processes of weighing. What we actually do when we weigh something is measure the force on the object produced by the gravity of the Earth. This force is directly related to the 'amount of matter' in the object (and in the Earth).

There are two ways of measuring this force. First we can *compare* the forces on an object to be weighed and on standard masses using a balance (Figure 1.17(a)). When the masses are equal, so are the forces on the two sides of the balance. Alternatively, using a previously calibrated spring, we could directly measure the force due to gravity on our unknown object (Figure 1.17(b)).

The distinction between force and mass becomes apparent if we think of taking both sets of weighing equipment to the Moon. Now the balance will still tell us that we have a kilogram of sugar. However, the spring will say there is only one sixth of a kilogram because the force due to gravity has changed. But it is the same bag of sugar containing the same 'amount of matter'; its mass has not changed. Its weight has changed, this being the force due to gravity. We will discuss units for force in later; clearly they are not the same as units of mass.

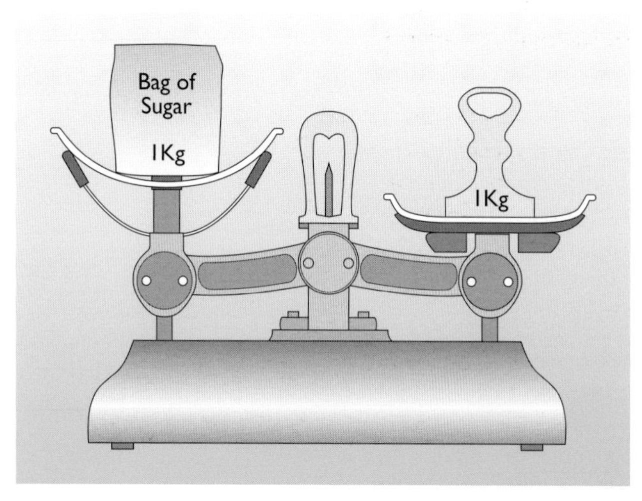

(a)

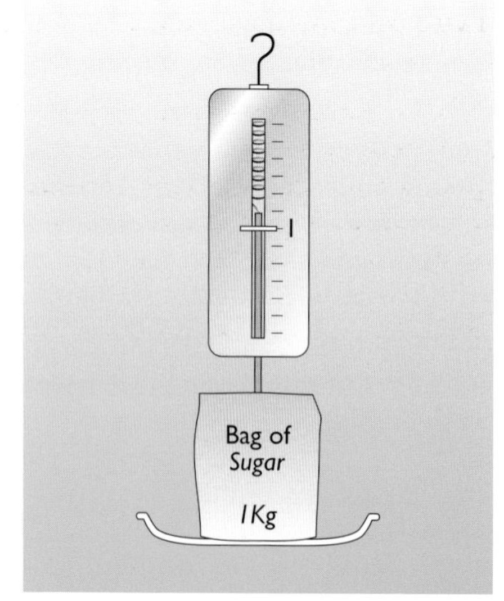

(b)

Figure 1.17

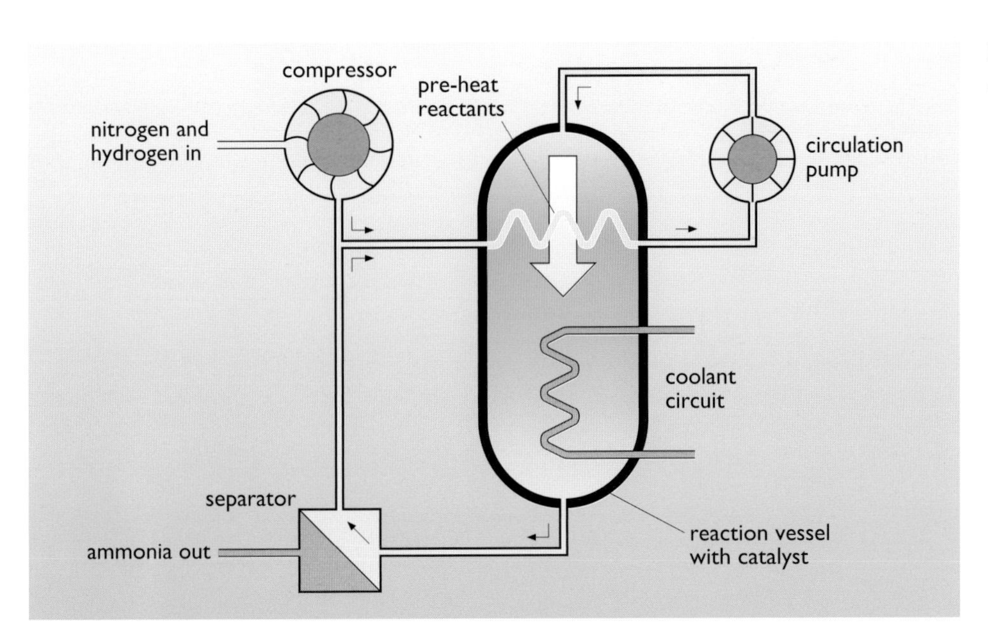

Figure 1.18
Ammonia production

Having collected the arguments from chemistry on the conditions to make ammonia it remains to ask how to make it *quickly*. Typically plant will only be economic at production rates of a few thousand tonnes per day. The optimum pressures and temperatures need to be found. That is a job for theory and for development research.

So, with all this science understood, our chemical engineers have to design and build a system that takes advantage of it. Figure 1.18 indicates the principles of the plant. The gases at high pressure are pumped around the circuit shown (follow the arrows starting at the circulation pump, which pumps the mixed, but unreacted, ingredients into the reaction vessel). The gases enter the reaction vessel at the top and some of the nitrogen and hydrogen react to form ammonia as they pass over a catalyst within the reaction vessel (see ▼Catalysts and converters▲). The mixture leaving the reaction vessel at the bottom passes to the separator where the ammonia product is removed. The un-reacted nitrogen and hydrogen, together with enough extra nitrogen and hydrogen injected from the compressor to make up for what has reacted, return to the pump via a heat exchanger within the reaction vessel which both pre-heats the reactants and removes some of the reaction heat. More excess heat is removed from the reaction vessel by the coolant circuit. Thus a continuous production of ammonia is achieved. I will not go into the subtleties of the internal design of the reaction vessel which make the process efficient, nor of the control systems which link the heat and gas flows. Rather I will concentrate on the engineering of the reaction vessel itself.

The ammonia reaction vessel is large, perhaps 20 m high by 2 m diameter, which is about 63 m^3 volume (see ▼Units of area and volume▲). For efficient performance, the pressure needs to be some 350 atmospheres, or in SI units, 35 MN m^{-2}, which is read as '35 meganewtons per square metre'. ('Pressure' is a way of relating a force to the area over which it is applied; force has units of newtons.)

▼Catalysts and converters▲

You will doubtless have come across the term *catalytic converter*, probably in the context of cars and exhaust emissions. A catalyst is something that speeds up or otherwise aids a chemical reaction. A true catalyst doesn't participate in the reaction at all. It just helps it on the way, and is unaffected itself, being left behind as good as new once the reaction is over. In cars, a chamber with a platinum-based catalyst is fitted as part of the exhaust system. Some gases in the exhaust, particularly nitrogen oxides and unburnt petrol molecules, decompose when they contact the catalyst into more benign gases.

In the case of ammonia production, iron granules with a large surface area are the usual catalyst.

▼Units of area and volume▲

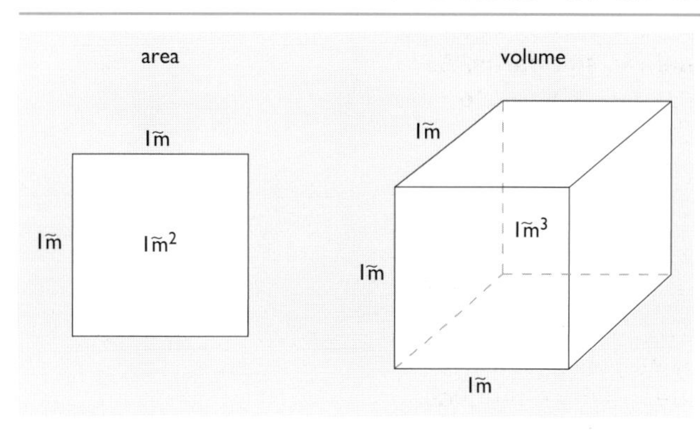

Figure 1.19

The basic units of area and volume are the 'square metre' and the 'cubic metre' (Figure 1.19). These are simply shapes which have sides that are 1 m in length.

Square metres are inconveniently small for measuring land holdings, even peasant land holdings, so the French revolutionary committee decided to define 100 m^2 (i.e. 10 m × 10 m) as 1 are (pronounced 'air'). The familiar land measure is the 'hectare'.

SAQ 1.8 (Learning outcome 1.1)

Use Table 1.2 to remind yourself what 'hect' means and then work out how many square metres there are in a hectare.

Cubic metres, on the other hand, are inconveniently large for measuring a person's daily intake of food or drink, for example, so a smaller unit of volume, the litre, was defined. A litre is the volume of a cube of 1 decimetre edge length. The symbol for 'litre' is its initial letter, l, but this is so easily confused with the number 1 (for example, eleven litres can be written as 11 l) that it is usually safest to write out the word litre.

An even smaller unit of volume, the millilitre, is also useful. It is one thousandth of a litre. The symbol for millilitre is ml, which is not so ambiguous as the litre symbol.

SAQ 1.9 (Learning outcome 1.1)

(a) Use Table 1.3 to remind yourself what 'deci' means, and then work out how many litres there are in a cubic metre.

(b) Wine bottles are usually labelled as having a capacity of 75 cl. How many ml is that?

(c) What is the relationship between 1 ml and 1 cm^3?

(d) The standard size of a British brick is 9 × 4.5 × 3 inches. First guess its volume in litres, then calculate it. (Take 1 inch to be 2.54 cm.)

(e) How many bricks make a cubic metre of wall (ignoring the mortar)?

The contents of the reaction vessel, at a temperature of several hundred degrees Celsius, include a poisonous gas, ammonia, and an explosive one, hydrogen. The consequences of a leak would clearly be dire. So, beyond the primary engineering to provide function, the issue here is *safety*. How do you make a vessel of that size, capable of holding the pressure, with at least six pipes going into it, and *guarantee* that it is safe? The answer lies largely in 'past practice'. Because pressure vessels have been made for a long time, for many purposes (for example, high-pressure steam boilers in power stations), the 'know how' has gradually developed. Each increment of extra performance demanded from time to time has pushed knowledge a bit further.

4.3 Standards

When boilers were first made for steam engines, little was known about steel. Plates were riveted together and engineers were ignorant of the stresses in the material. Of course there were successes and there were failures; both can be learnt from, even if the lessons are only 'use thick steel plates'. Later it became possible to weld plates, and new lessons had to be learned.

All this accumulated experience is brought together into a 'standard'. The standard will dictate critical features of the design, construction methods and safety testing of pressure vessels. For example it will guide the designer in relating wall thickness to diameter and pressure. For the ammonia vessel I described, the plate would have to be 150 mm thick. Then permitted types of steel and the welding methods will be specified. Finally the whole thing has to be tested to beyond its working pressure.

Standards govern the design and construction of virtually everything which carries any safety implications. Woe betide the company which transgresses the standards and whose product is then responsible for an accident. We shall discuss engineering with standards in more detail later in the course. For the moment though it is interesting to contemplate the apparent conflict between opportunities for innovation and the requirement to adhere to standards. As standards derive from *past* practice, they are apparently not conducive to innovation. What is the way out of this dilemma? In reality standards are revised, amended, and superseded to account for changes in knowledge and practice.

You can look around your home to try to find evidence of this – you should find the phrase 'Conforms to BS *xxxx*' (the BS standing for 'British Standard', and the *xxxx* being a number) or similar. Electrical goods, or their packaging or instruction manuals, would be a good place to search. For, example, I found in the instructions for a French-made electric kettle that its moulded-onto-the-wire mains plug conformed to BS 1363 and that only fuses approved to BS 1362 should be used as replacements. Furthermore the kettle was constructed to comply with the radio interference requirements of EEC directive 87/308/EEC. I have also noticed the number BS EN 228 (where the EN indicates that the standard is applicable throughout Europe, not just in the UK) on petrol pumps for unleaded fuel. These last two regulations appear to control the function of the product rather than its safety.

Exercise 1.7

Which of the following list would you put into the categories of one-off production, mass production, and bulk production?

(a) A modern 3-bedroom house on a development of 30 houses.

(b) Liquid oxygen for a chemical laboratory.

(c) A computer floppy disk.

(d) A music CD.

(e) The Channel Tunnel.

(f) A fax machine.

(g) Garden compost.

(h) Car tyres.

(i) A hand-made paperweight.

4.4 Summary

So where have we reached in this part of Block 1?

First we have seen that, beyond the shelf-building level of engineering, which I suppose most of us can undertake with more or less skill, there is a profession of engineering which dates back a long way. We shall examine the origins of this profession in the next part. Professional engineers, working either alone or in teams, can undertake much more complex tasks by virtue of their talents, training, knowledge and experience. The great diversity of engineered products has led to many specialisms within the profession.

Secondly, with each of the short case studies it has become apparent that the engineering of any time rests on prior achievements. The ballpoint pen, by virtue both of its materials and the construction methods could not possibly have been made by the engineers of the Pont du Gard, despite their great skill and expertise. Clearly the builders of that lovely bridge nevertheless drew upon the prior knowledge of how to construct arches. This aspect of the 'progress' of engineering will occupy our attention later.

The third aspect of engineering we have seen in action is its organizational aspect. One of the engineer's roles is to ensure that resources are available to achieve the current goal. Engineers control the actions of others to bring about the success of a project: they depend on the skills of their stone masons or lathe operators or welders. Badly organized engineering is just as chaotic as a badly directed orchestra.

A fourth important realization is that engineering is *quantitative*. Measurement and calculation are vital to its practice. At its simplest it may only be spatial quantification, as for example to build the bridge. Further down the line, though, a panoply of measurement capability is demanded. Ammonia synthesis calls for measurements of temperature, pressure, gas flow rates and chemical concentrations even within the small part of the process that we have looked at. Notice, however, that there is a big difference between being *quantitative* and being *scientific*. The builders of the Pont du Gard had no scientific knowledge of, say, the stresses in their structure; they did not design it scientifically, using theories of how arches worked. In contrast, the ammonia pressure vessel was constructed with full knowledge of how the material *should* behave, even to the extent of knowing the consequences of there being a flaw (in the welds of the vessel, for example).

Measurement implies units to measure in: metres of length, kilograms of mass and so on. Historically there have been many definitions of standards for measurement. Now we have settled more or less completely on the SI (*Système International)* which has being described in earlier sections. But the SI is not yet universal: American engineers flew to the Moon using feet and inches for length, and pounds for mass (see ▼Different systems of units▲).

Not every engineering decision about a length needs to be based on an actual measurement. Most of Blanc's gauges for the parts of the M1777 musket lock were of the 'go/no go' type; a particular pin, for example, had to be small enough to enter one gauge hole but big enough not to go into another. So the pin diameter lay between the two hole diameters. The alternative approach of defining the size of the pin as, say, $5.00 \text{ mm} \pm 0.01 \text{ mm}$ and providing an instrument to measure the pin to that accuracy was not available to Blanc – neither the micrometer nor the millimetre had by then been invented. For his purpose, all the pins had to be the same; give or take a bit. It probably did not matter *exactly* what that size was.

Exercise 1.8

When you place a ruler alongside another rule, or tape measure, do the lengths match up *exactly*? If there is a difference, is it significant?

We have also seen an example of how science can inform engineers of how to proceed. Without a thorough understanding of ammonia chemistry on the part of engineers, the Haber–Bosch ammonia manufacture process could never have been developed. But, as we shall see, some remarkable achievements in materials processing *have* been developed without scientific understanding. We speak of the Bronze Age or of the Iron Age in archaeology because the people of those times had discovered how to get metals from minerals; and in the case of the Bronze Age, how to make various copper alloys (not just

▼Different systems of units▲

The SI system of units is not the only one which is available, or indeed, in common usage. The most widely used alternative is the system of pounds for mass, and feet for length. This is the 'Imperial' system of units. A table comparing these two systems is given in an Appendix to this part of the first block.

In the US, Imperial measurements are more common than the SI metric system. This can lead to problems if there is a mix-up as to which units are being used. The NASA Mars atmosphere probe was lost in mid 1999 due to a mix up between the two systems. A computer directed the wrong thrust to be applied to bring the craft into Mars orbit because its program was in one system and its data in the other. Interestingly, Americans refer to the Imperial system as 'English units'.

bronze) so as to enhance the properties of their metal. They did these things without any knowledge of chemistry. The role of science for engineering has been largely two-fold:

- to provide new perceptions of what might be done – electric motors, atom bombs, transistors, radio, etc.;
- to enable developments to be made more effectively and efficiently.

Given a decent theory of motion, for example, it is easy to calculate how powerful an engine should be installed in a 40 tonne truck to enable it to climb a one-in-twenty hill at 90 k.p.h., and any other performance criteria that might be imposed on the design.

This brings me to the final and most important item of this summary – *design*, which lies at the heart of all professional engineering. Whatever enterprise engineers are going to co-ordinate, it starts off as a mere feeling. A 'wouldn't it be nice if?' sort of thing. Along the line someone has to turn that sort of vague aim into a *practicable* idea. By practicable I mean not only that whatever has been conceived of will meet the function implied by the goal, but that it can be made too. For how many ages have people watched birds and longed to fly? In mythology, Daedalus (unquestionably an engineer) made wings of feathers glued on with wax, enabling him and his son, Icarus, to fly. But Icarus flew too close to the Sun and the wax melted, resulting in the first fatal air crash. Of course, this didn't really happen, nor could it have happened. We now realize that human muscle power cannot be sufficient to get us off the ground by flapping wings. Leonardo da Vinci drew pictures of a screw-driven machine (Figure 1.1). But he had no way of powering it. The Montgolfier brothers became airborne in 1783 in the first hot-air balloon, but that's not real flying, as the balloonist is at the mercy of the wind. Eventually, when the engine was available we made it – less than 100 years ago. The development from the Wright brothers' contraption to a modern aircraft demonstrates again each engineer's debt to those who have gone before.

Figure 1.20 Flying machines, ancient and modern

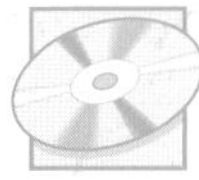

A gallery of images showing the development of flying machines can be found on the T173 CD-ROM.

The point I want to make is that design is the imaginative, creative part of engineering. It is the *sine qua non* ('without which, nothing') of engineering, and it comes from inside our heads. It is a manifestation of imagination – putting experiences and ideas together in new ways to conceive of new possibilities. Note the plural; there is usually more than one way of reaching the goal.

One of the odd (and fascinating) things about imagination is the surprises it springs. You can't forecast what is going to come in response to any particular stimulus. You can't necessarily even spot that there has *been* a stimulus. People had been pushing heavy things around on rollers for thousands of years without the flash of inspiration that invented the wheel. The idea of fixing an axle to the load and threading each end *through* rollers so they stay with the load is a subtle perception: a beautiful example of an imaginative surprise. It could have happened in any of thousands of minds at any time over thousands of years, or maybe it just happened in one mind. Whichever it was, no doubt it was accompanied by the thought 'Why didn't I think of that before?'.

Well, I presume something like that happened. But the idea had to be followed by action. Though the inventor could perceive a wheel in the mind's eye, none yet existed. *How* to put the idea into practice would then call for more imagination. This time it is a different kind of imagination; ideas can be tested straightaway against the yardstick of practicability.

5 Learning outcomes

When you have studied this part of the block you should:

1.1 Be familiar with (and be able to use) SI units of time, length, mass, area and volume.

1.2 Describe what distinguishes human capabilities from those of other animals.

1.3 Describe examples of 'intuitive' engineering, recognizing variety within the design domain, the opportunistic use of materials and the role of development.

1.4 Recognize the contexts of need, current practice and availability of materials and manufacturing ability in producing a successful invention.

1.5 Be aware of some of the key steps in our technological development.

1.6 Recognize engineering as a creative approach to problem solving in the practical domain.

Appendix Metric/Imperial conversion table

Imperial to metric

	Imperial		*Metric*
Length	1 mile	≈	1.61 km
	1 furlong	≈	200 m
	1 yard	≈	914 cm
	1 foot	≈	30.5 cm
	1 inch	≈	25.4 mm
Area	1 acre	≈	0.405 ha
	1 square yard	≈	0.836 m^2
Volume	1 gallon	≈	4.55 litres
	1 pint	≈	568 ml
Mass	1 ton	≈	1.02 tonne
	1 pound	≈	454 g
	1 ounce	≈	28.4 g

Metric to imperial

	Metric		*Imperial*
Length	1 km	≈	0.621 mile (about 5/8 mile)
	1 m	≈	1.09 yard
	1 cm	≈	0.394 in
	1 mm	≈	39.4 thou
	1 μm	≈	0.000 039 4 in
Area	1 km^2	≈	0.386 sq. mile
	1 ha	≈	2.47 acre
	1 m^2	≈	1.20 sq. yard
	1 cm^2	≈	0.155 sq. in
Volume	1 m^3	≈	220 gallons
	1 litre	≈	1.76 pints
Mass	1 t	≈	0.984 ton
	1 kg	≈	2.21 lb

Answers to exercises

Exercise 1.1

Although some of the early hominids had tool-making abilities, we find a marked development in tool making about 2 million years ago in the *Homo* branch.

Exercise 1.2

The jet travels 800 km in 3600 s, so to travel 1 km takes

$(3600 \div 800)$ s = 4.5 s

So the airliner takes 4.5 s to travel 1 km.

Exercise 1.3

The engineer will have to find finance from somewhere, so that brings in the bank manager. The bank may want to see a business plan before stumping up the cash, so market research may be needed. The engineer may want to patent the invention, so that will involve patent agents. In order to sell the product, there will have to be advertising and sales people. The engineer may need advice from other specialists in the manufacture of the invention. All this is a long way from the DIY approach of the first bow and arrow!

Exercise 1.4

Weight and bulk are no longer a problem. You'll need a shelter but it doesn't have to be portable. Stone and large timbers become feasible building materials, and buildings can be bigger and permanent. Grain will have to be stored – that means a dry place. To cultivate any reasonable area of ground, some kind of plough should be invented and it would be a good idea to recruit some sort of large docile animal to pull it. We could domesticate some animals for meat as well as for draught purposes.

Exercise 1.5

The exceptions are steel (a mixture of iron, carbon and other elements), glass (usually based on silicon and oxygen), leather (a natural material containing carbon, nitrogen, oxygen and hydrogen, with many other elements in small quantities), and plastic (a generic name for a host of materials based on carbon and hydrogen, often with other elements also).

Exercise 1.6

Supply of vegetable oil, which requires farmers to grow oil seed;
oil processing plant;
harvester for crop;
plough to cultivate land;
fertilizer and pesticides;
roads to transport to and from farm;
bricks/cement to build factory;
people to make all these things;
fuel for tractors, trucks, etc;
steel industry, which implies mining;
alkali industry;
soap-making plant;
machine to press soap into bar;

paper to wrap bar in;
printing for wrapper – inks, dyes;
cardboard boxes to pack bars for transport to shops;
forestry for woodpulp for packaging;
electricity supplies for factory machines.

My three minutes are up and I've not got the soap to my home yet!

Exercise 1.7

(a) A modern 3-bedroom house on a development of 30 houses. This is a one-off (no economy of scale here).
(b) Liquid oxygen for a chemical laboratory. This used bulk production.
(c) A computer floppy disk. This is mass production.
(d) A music CD. Mass production again.
(e) The Channel Tunnel. One-off.
(f) A fax machine. Mass production.
(g) Garden compost. Bulk production.
(h) Car tyres. Mass production.
(i) A hand-made paperweight. One-off.

Exercise 1.8

I found that there was a slight difference between the two rulers that I chose. However, since I only ever use them for coarse measuring (to within 2 mm or so) this is not a problem.

Answers to self-assessment questions

SAQ 1.1

I came up with these answers:

- mining of the Earth's crust to extract mineral and other resources;
- large-scale burning of fossil fuels which has generated large amounts of carbon dioxide and raised the temperature of the atmosphere (if you accept the current theory of global warming);
- agriculture and city-building (two examples, really), both of which have dramatically altered the lands from their original natural states.

You may have come up with other answers to the question: I'm sure that they are equally valid.

SAQ 1.2

(a) In a year of 365 days, the number of 'French' seconds would have been

$100\,000 \times 365 = 36\,500\,000$

(b) Again assuming 365 days in the year, the number of seconds is

$365 \times 24 \times 60 \times 60 = 31\,536\,000$.

If you took the number of days to be 365.25 or 366, then your answer should have come out as 31 557 600 or 31 622 400 respectively.

There is a difference of 14 per cent between the current system and that proposed by the French Revolutionary Committee. That is, the proposed 'French' second would have been 14 per cent shorter.

SAQ 1.3

The longer the stick, the more easily it will bend; the thicker the stick, the more force it will give. I need to be able to carry the bow while I'm out hunting; it should not be too heavy – long and thick means heavy. If it's too long it will be clumsy to handle; too thick and I shan't be able to bend it. But getting a decent force is important: the heavier, faster and sharper the arrow the more damaging it will be to the prey.

So, from the using viewpoint I'll want a stick as long as I can conveniently handle and as thick as I can bend without over-straining my body. A straight stem of green hazel would be nice and springy. A pole about as long as I'm tall and as thick as my thumb feels about right. Can I find a piece of rawhide or twisted fibre string (we've got the know-how for both) long enough for such a stick? I'll need to notch the stick to stop the string sliding off the bow when I draw it. I'll experiment with arrows of various lengths and weights to see what goes best.

SAQ 1.4

In this case, the available materials and manufacturing skills were no problem: hence the product could be constructed and marketed. Current practice when the vehicle was released (around 1984) was to travel either by car, foot or bicycle. The vehicle was smaller and cheaper than a car, and cheaper to run, but also slower and had a lesser range. The main thing it fell down on was the context of need: it was not sufficiently appealing as an alternative to either car or bicycle for people to rush out and buy one. Its low height also meant that the driver could feel threatened in heavy traffic.

SAQ 1.5

(a) *H. sapiens sapiens* has been the only surviving species of *Homo sapiens* for 30 000 years. We have been city dwelling for 5000 years. So, of the time during which we have been the surviving species of *Homo sapiens*, the percentage of the time for which we have been city dwellers is:

$(5000 \div 30\,000) \times 100\% \approx 17\%$

(b) We have been industrialized for 200 years. So, of the time during which we have been city dwellers, the percentage of the time for which we have been industrialized is about:

$(200 \div 5000) \times 100\% \approx 4\%$

The time during which we have been city dwellers is about 0.7% of the time we have been the surviving species of *Homo sapiens*.

(c) The *Homo* genus, in one species or another, has been on earth for 2 000 000 years. We have had electricity for 80 years. As a percentage, this is:

$(80 \div 2\,000\,000) \times 100\% \approx 0.004\%$.

This gives a useful perspective on how modern our engineered world really is.

SAQ 1.6

(a) (i) 2000 mm is 2.000 m, which is greater than 1.8 m.
(ii) 1 mm is 1/1000 m whereas 1 km is 1000 m, so the kilometre is the greater.
(iii) 100 μm and 0.1 mm are the same. 100 μm is 100×10^{-6}, or 10^{-4} m, and 0.1 mm is $0.1 \times 10^{-3} = 10^{-4}$ m.

(b) 1 mm is $\frac{1}{25.4}$ of an inch, so 1 m is $\frac{1000}{25.4}$ inches, which is 39.37 inches

(c) The magnification is:

$$\frac{2\text{ cm}}{2\ \mu\text{m}} = \frac{2\times10^{-2}}{2\times10^{-6}} = \frac{10^6}{10^2} = 10^4 = 10\,000$$

SAQ 1.7

This is what I came up with.

- The ball must fit sufficiently tightly that the ink doesn't ooze out. On the other hand, the ink must not be completely blocked. Also, the ball must not be damaged during the assembly process. (Maybe that counts as three problems, rather than one.)
- The ink must not be too fluid or it will leak from the pen. (In many pens the end of the ink holder is open; you may have carried such a pen in a pocket, and discovered that the ink becomes more fluid at body temperature, leading to a trip to the dry-cleaners.)
- The join between the metal top and the ink holder must be a good seal.

You may have come up with equally valid problems which need to be addressed for the product to work successfully.

SAQ 1.8

'Hect' means ×100, so 1 hectare is 100 are, which is

$100 \times 100\ \text{m}^2 = 10\,000\ \text{m}^2$.

SAQ 1.9

(a) The 'deci' prefix means ÷10, so there are 10 dm in one metre. Hence,

$1\ \text{m}^3 = 10 \times 10 \times 10\ \text{dm}^3 = 1000$ litres

(b) The 'centi' prefix means $\div 100$, so

$75 \text{ cl} = (75 \div 100)$ litres

$= (75 \div 100) \times 1000$ ml

$= 750$ ml

(c) $10 \text{ cm} = 1 \text{ dm}$ so

$10 \times 10 \times 10 \text{ cm}^3 = 1 \text{ dm}^3 = 1$ litre

So 1000 cm^3 make 1 litre, and hence 1 ml is the same as 1 cm^3.

(d) The volume of the brick in cubic inches is

$9 \times 4.5 \times 3$ cubic inches $= 121.5$ cubic inches

There are 2.54 cm in an inch, so $(2.54)^3 \text{ cm}^3$ make 1 cubic inch. So the volume of the brick is

$121.5 \times (2.54)^3 \text{ cm}^3$

$= 121.5 \times 16.4 \text{ cm}^3$

$= 1992$ ml

$= 1.992$ litres (or very nearly 2 litres)

My guess, thinking of a litre box of drink by comparison, was that the brick was 'similar' and would have a volume of about one litre. I was wrong by a factor of 2.

(e) There are 1000 litres in a cubic metre. We know from (d) that each brick is about 2 litres in volume, so we'll need approximately 500 bricks. (This sort of approximate calculation is typical of the way engineers work, often as a prelude to more exact calculation.)

Actually, each brick is 8 ml short of being 2 litres, so 500 of them will be deficient by 500×8 ml, or 4 litres, which is slightly more than two bricks' worth (each brick being just under 2 litres in volume). So 503 bricks will meet the requirement.

References

Alder, K. (1999) : *Engineering the Revolution,* Princeton University Press.

Clark, G. (1961) *The Dawn of Civilisation*, Thames & Hudson, London.

Wenke, R. J. (1999) *Patterns of Prehistory: Humankind's First Three Million Years*, Oxford University Press.

Part 2
Why we do ‘engineering’

T173 Course Team

Academic Staff

Dr Michael Fitzpatrick (Course Team Chair)
Graham Weaver
Dr Nicholas Braithwaite
Adrian Demaid
Dr Bill Kennedy
James Moffatt
Dr George Weidmann
Mark Endean
Jim Flood
Dr Suresh Nesaratnam
Professor Bill Plumbridge

Consultant

Shelagh Lewis

Production Staff

Sylvan Bentley (Picture Research)
Philippa Broadbent (Materials Procurement)
Daphne Cross (Materials Procurement)
Tony Duggan (Project Control)
Elsie Frost (Course Team Secretary)
Andy Harding (Course Manager)
Richard Hoyle (Designer)
Allan Jones (Editor)
Howard Twiner (Graphic Artist)

Contents

6 Introduction to Part 2

In the first part of this block, we looked at the historical development of engineering, from the sort of engineering that can be undertaken at an intuitive level, through to large-scale modern projects which require precise understanding of materials behaviour, financial planning and careful co-ordination of innumerable specialist professional engineers and craft workers.

We are now going to move on to look at the practice of engineering in more detail. We'll start by continuing our examination of what drives engineering, and then begin looking at some of the tools that are used in engineering; by which I mean intellectual and resource tools, rather than shovels and rulers.

As a reminder, here are the aims of this block.

- To introduce the course by defining what engineering is, and the social and economic contexts of its practice.
- To reveal the origins and landmarks of engineering.
- To show engineering as a creative and intellectual human activity.
- To explore why and how we engineer.

7 Why do we do engineering?

7.1 The ancestry of invention

One of the messages from Part A is that we engineer because we can't help it. We always try to find ways to solve problems in our environment: how to transport water to a city; how to make money from a disposable pen; how to make explosives without nitrates. A proverb, perhaps expressing a tendency to laziness, has it that we need a spur: 'necessity is the mother of invention'. Certainly, invention is a first step of engineering. But I don't think that bit of wisdom delves deep enough. Our ultra-brief skim through our distant ancestry has revealed that only within the final 1 per cent of humans' time on the planet has there been much inventiveness around. Necessity, in the sense of survival, was apparently satisfied without even bows and arrows, let alone cars and transistors, for a long, long time. Before the spur of *necessity* must come the *ability* to invent, a quality we have identified as innate to *homo sapiens sapiens*. Just as it is the scared bird that flies which is the survivor, so it is in our genes to set up imaginative responses to problems or difficulties we experience in our environment. This, coupled with our other bit of evolutionary good fortune, our dextrous hands which enable us to construct things, has made us very adaptable. So we don't always see problems or difficulties so much as opportunities, and opportunity is also a parent of invention. Finally, we are very curious creatures. We notice things, like sharp, split stones or springy sticks, how porcupines behave, and which rocks make copper when heated on a fire. All this information feeds our imagination – and becomes part of the culture to be handed on through the generations. Curiosity then, is also a prerequisite to inventiveness, or at least, is a tool for doing invention. And, if all this seems relevant only to the distant past, think of the software explosion that has been the main topic of invention since about 1980 – it's still the same set of drivers at work.

Moreover, these same qualities are at the root of the continuously accelerating growth of engineering. Curiosity has led us to science, which has provided a greatly improved understanding of what is possible. In the field of electricity, for instance, where intuition is of little help, we are completely dependent on scientific insight: you can't judge whether a live wire is safe to touch in the same way that you can judge whether a chair is strong enough to sit on.

Opportunity and ability develop from previous achievements. Steam engines, as sources of increased power, allowed the development of railways, factories, big ships and deep mines. Internal combustion engines (see ▼Internal and external combustion▲), with greater power-to-weight ratio and using convenient fluid fuels, have let us have the motor car and made powered flight possible. The ever increasing repertoire of materials (reinforced concrete, plastics, semiconductors, aluminium, radio-active isotopes, carbon-fibre composites, etc.) stimulates fresh ideas for products and for ways of making them. The present computer-based revolution is opening hitherto undreamt of possibilities in design calculations and in process control. But we'll think more about these things later.

▼Internal and external combustion▲

You may often have heard the term 'internal combustion engine' applied to the engine of a car. The term means what it says: the fuel is burned internally. For a car engine, that means inside a sealed chamber – a cylinder within the engine itself. An external combustion engine would be one where the fuel was burned outside the engine. The railway steam engine is a good example of external combustion: the coal is burned in a firebox to heat the water in the boiler, from which steam is conveyed to the piston. In internal combustion, the energy released by the burning fuel can be harnessed more efficiently: it isn't transferred first to another medium (such as the water in the steam engine).

The more pragmatic view of why we engineer is to be found in looking at the *purposes* which engineering satisfies. Right at the start we imagined engineers describing their activity by its purposes – to make functional things, to generate profit for a company or wealth for a country, to improve standards of living. Already you can see diversity of purposes, indeed a hierarchy of purposes, since making functional things may be a route to generating profits for your company, which in turn represents part of the wealth of the country. To study links like this, it will pay to find out how control over what is to be engineered is exercised. As so often, it all started off very simply and has grown to confusion as time has passed.

7.2 Engineering for function

The simplest, most straightforward reason for engineering is to make something which provides a *function*. The Roman aqueduct or a plant for making ammonia sit just as happily within that generalization as a split stone conceived and made to serve as a scraper. What makes the later examples different from the stone tool are the complexity of the chains of actions and the repertoire of starting points. In there lies a distinction between 'craft' and 'engineering' which I need to explore.

Activity 4 Engineering and craft

The fourth audio track looks at how engineering for function has developed from the stage where engineering was an individual activity, performed by intuition and ritual, through the stage where groups of people began to specialize in offering services of one specific product, through to the mass-production systems of today, where individual efforts are valued for craftsmanship rather than for advanced engineering skills. You can either break now to listen to this track, or return to it at a more convenient time. Remember to re-visit SAQ 2.1, though.

SAQ 2.1 (Learning outcome 2.1)

Looking around you, consider which of the objects in your surroundings have been crafted rather than engineered?

By the eighteenth and early nineteenth centuries, it was becoming clear that there was a distinction between the jobs carried out by craft workers and those carried out by engineers. Remember from Part A how the engineers' plans for interchangeable parts for musket locks were shelved because the locksmiths (craftsmen) were seen as so strategically important that it was unwise to offend them. And from 1811 to 1816 gangs of 'Luddites' in the Midlands and north of England smashed engineered factory machinery in order to try to protect their craft livelihoods.

Nowadays 'craft' products are valued chiefly for the fact that they *are* hand made. For something special we like the idea that someone's soul has been involved with its making. But for supplying everyday functions, we expect that machines and engineers have been at work. Mass production, with its attendant necessary accuracy, makes things cheaply. It is labour which is expensive: capital invested in machinery gives a better return – provided, of course, that you make what the market will buy. And equally, of course, market expectations change.

As you well know, the appliance of science has continued to open opportunities for functional products. At the time of writing, one of the television channels had recently reconstructed a middle-class home of 1900 as exactly as it could, and installed a family to live in it for three months. The family was none too happy! There was no domestic electric power, so no telephones, television, radio, refrigerator, vacuum cleaner, washing machine, dishwasher, home computer. The facilities that were available were seen as inconvenient, even dirty: heating and cooking by coal, lighting by gas, limited hot water. There were no plastic artefacts, no aerosols, no frozen foods, no artificial fibre clothing and, of course, no car.

SAQ 2.2 (Learning outcome 2.2)

Think of six more things, domestic or other, which are now available which were not when you were a child.

Let's return to our story of engineering for function. A most important characteristic of a functional product is that it should actually do its intended job: in legalistic language it must be 'fit for function'. Horseless carriages have been around now for about a hundred years and have changed markedly from the early models (Figure 2.1). Apparently the veteran car in Figure 2.1 was 'fit for function', but it would not live up to our present expectations. It would have been cold, draughty, noisy, rattly, and it would not have either gone fast or stopped quickly; and it probably only went approximately where you pointed it. Evidently our expectations of performance have increased. The first thing this example shows is that functional products *develop*. The first engineering solution is rarely if ever the final word. Why?

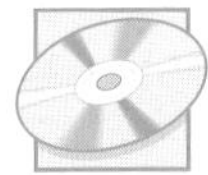

A gallery of images showing the development of the car can be found on the T173 CD-ROM.

Figure 2.1 Veteran car

One important answer to this interesting question is that for virtually any function there is a near infinite number of design possibilities. For several reasons, the earliest products that meet the function have explored only a tiny fraction of this 'design space'. Reasons typically include the following.

(a) The function has as yet been only narrowly defined in terms of performance.

(b) Performance possibilities/desires are revealed but not realized by the early versions.

(c) The new product depends on production methods developed for other devices. It might be possible to improve these methods specifically for the new function.

(d) As new materials and techniques become available, different designs become possible.

(e) Until the new function has become popular, it will be difficult to attract investment.

All these apply to the development of cars. The earliest cars perhaps merely showed that a horseless carriage powered by an internal combustion engine was possible. The engine itself and the transmission (the gears and drive shaft to transmit the motion from the engine to the wheels) were the new technology which required the most attention from the engineering team. Coachwork, suspension, wheels, braking and steering mechanisms could be inherited with modification from the existing technology of horse-drawn vehicles. Personal transport was at that time only available for the rich, so the new vehicles had only a small market, limiting production and hence keeping unit costs high.

The military imperatives of the First World War did much to promote development of more powerful and reliable engines and faster methods of sheet-metal processing for bodywork, by using machine pressing instead of panel-beating by hand. When the function of a product is turned towards military use, the money (and hence the engineering expertise) available for developing the design of the product suddenly becomes much greater. Many of the developments so produced find use in ordinary products for the consumer market – from tanks and military transport to cars, for example.

When Henry Ford reformulated the specification of the car in terms of popular availability rather than performance, he not only had to design a car which was utilitarian rather than luxurious, but also a production system that would hack unit costs down and down. Mass production of parts and rapid, high-volume processes in assembly comprised the route to low cost, and hence to a vastly expanded market. For just one example of the impact of new materials in cars, it has been said that, if it were not for the fact that plastics have largely replaced leather for car upholstery, beefsteak would now be a by-product of the automotive industry.

You can find another answer to the question of why engineering products develop over time by thinking about the car in Figure 2.2, the ill-starred Trabant manufactured in East Germany throughout the years of communist rule. This car did *not* develop, even though it never was much good; it was barely 'fit for function'. The reason Trabants never changed or improved was that the environment of its production put no pressure on the manufacturers to make any changes or improvements. The factory produced the quota it was

Figure 2.2 Trabant

▼Combining measurements: speed, velocity and acceleration▲

So far in our asides on measurement, I've introduced fairly fundamental quantities: mass (weight), length and time. In fact, the units which describe these three quantities (kg, m and s), along with a unit for electric current (the ampère, A), form a basis by which virtually everything can be measured. In practice a combination of units is often needed, which can be quite complex. To get round this complexity, a new unit has often been defined for measuring a particular effect or phenomenon – even though in principle the unit can be boiled down to a mixture of those mentioned above.

When it comes to cars, the main criteria of performance relate to how a car moves. To describe things like how fast a car will go or how it accelerates calls for units of speed and acceleration. These units are constructed by combining units of distance and time in appropriate ways. This should be fairly straightforward to understand: we think of car performance in terms of how long it takes us to reach a destination (traffic jams permitting), and how well the car responds to a push on the accelerator pedal.

To get at the idea let's start with the very familiar idea of 'speed' – measured, as you know, in 'miles per hour' for cars in Britain. This links units of distance (miles) and time (hours) with the word 'per' meaning 'in each'. So, if a journey of sixty miles has taken two hours you will recognize that (on average) in each hour a distance of thirty miles has been covered: that's thirty miles per hour. Obviously you *divide* the distance by the time to get the speed.

Miles per hour are not SI units however. We need to make a speed unit out of the SI units of length and time.

Exercise 2.1 (revision)

What are the SI units for length and time?

So in SI units, speed is measured in metres per second. The units of speed are therefore metres divided by seconds. We tend to say 'metres per second' rather than 'divided by'. If we write out the symbols as a formula, the units of speed are:

m/s

where the / symbol is used instead of ÷ to indicate 'divided by' or 'per'.

Another notation for 'per', which is particularly convenient when the units are expressed in symbols rather than words, is the use a negative power, or index. Thus metres per second becomes m s^{-1}. If you are not happy as to why, review the Maths Help section on Powers and Roots, pp. 353–5 of the *Sciences Good Study Guide*. (Just for added confusion 'per' turns up as the letter 'p' in 'm.p.h.' or 'k.p.h.' for speeds in miles or kilometres *per* hour, although this isn't strictly a correct scientific form of notation.)

SAQ 2.3 (Learning outcome 2.5)

What is the speed of a vehicle which travels 2 km in 2 minutes? Express your answer in metres per second.

Velocity

Although the words 'speed' and 'velocity' are used synonymously in general parlance, in science and mathematics there is actually a subtle distinction between them. Velocity contains the idea of *direction*. If the sixty mile trip was supposed to take you from Southampton to Brighton, it would have been important to head in the right direction. The *speed* would be 30 m.p.h. whichever direction, but the velocity would have been 30 m.p.h. due east. Quantities which contain direction within them are called *vectors*. Speed is the size (that is, the 'magnitude') of a velocity vector.

Acceleration

Acceleration is to do with how fast velocity changes. Just as velocity is a vector, so is acceleration. If you accelerate to avoid a charging bull, it rather matters which direction you take off in.

In general parlance, however, when people use the term 'acceleration' they are often refer to what scientists and mathematicians would call the magnitude of the acceleration, that is, the size of the acceleration. Most of the time this lax usage causes no problems.

A magazine article might claim that a sports car had 'an acceleration of nought to sixty miles per hour in five seconds', which is telling you something about how quickly the speed changes. A small economy car might take twelve seconds to achieve the same speed. The sports car has greater *acceleration*. The smaller car may be able to reach the same speed, but it will take longer. (Note that acceleration can be negative, which means that the speed is decreasing rather than increasing.)

In mathematics and science, acceleration is strictly defined as the change in velocity divided by the time during which the velocity changes. However, the *magnitude* of the acceleration is just the change of speed divided by the time. So what is the magnitude of the acceleration of the sports car going from 0 to 60 m.p.h. in 5 seconds? Evidently 60 m.p.h. ÷ 5 seconds, or an increase of 12 miles *per* hour *per* second.

Again, our answer is not an SI unit. To work out the magnitude of an acceleration we need the rate of change of the SI unit of speed. This means that in SI, the magnitude of acceleration is measured in metres *per* second *per* second. The unit of time appears twice. You might write the unit is m/s/s.

We can treat m/s/s as a double division, and manipulate the unit symbols by the same rules that we would use for manipulating algebraic variables. So

$$\begin{aligned} m/s/s &= (m \div s) \div s, \\ &= (m\ s^{-1}) \div s, \\ &= m\ s^{-2}\ . \end{aligned}$$

You sometimes hear this unit expressed as 'metres per second per second', which is the same as 'metres per second squared'.

SAQ 2.4 (Learning outcome 2.5)

(a) A car joins a motorway slip road, and accelerates from 15 m s^{-1} (about 34 m.p.h.) to 30 m s^{-1} (about 68 m.p.h.) in 8 seconds. What is the magnitude of the acceleration of the car in SI units?

(b) An athlete runs an 800 m race in 150 seconds. What is the average speed of the athlete during the race?

I mentioned earlier that in science and mathematics, acceleration is strictly defined as the rate of change of *velocity*. One consequence of this is that if something moves at constant speed, but changes direction (for example a satellite orbiting the Earth), then it is accelerating as far as mathematicians and scientists are concerned.

told to make (I suppose); their customers had only a limited choice of brands, most of which were equally uninspiring; the engineers had no incentive to make improvements; and any suggestions for investment would probably not have been approved by the people controlling their economy, for whom customer satisfaction was scarcely a priority.

Today's cars tend to be sold on looks and comfort, and possibly fuel economy, rather than speed or acceleration (see **▼Combining measurements: speed, velocity and acceleration▲**): the developments in design have reached a stage where a basic adequate level of performance is more than taken for granted.

7.3 Engineering for patrons

7.3.1 Introduction

The 'complexity' of society that we are recognizing as a hallmark of civilization necessarily contains institutions of political power. The idea of a city drawing its sustenance from its hinterland in return for specialized products of the city implies levels of control. Some system of government must emerge with power to direct the actions of the populace. No doubt the complex government of a city state would have evolved continuously and in an *ad hoc* manner as the scale grew from village to city. Whatever the governmental system, patronage of engineering projects was part of its activity. Immediately we can perceive a number of purposes for engineering: social, military, commercial, prestige.

Activity 5 Patronage of engineering

You should listen now (or when convenient) to the fifth audio track, on sources of patronage for engineering. This audio track gives an introduction to some of the areas of patronage which give a lead to engineering projects.

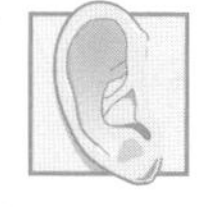

SAQ 2.5 (Learning outcome 2.3)

Consider the following items, and decide for yourself who was the 'patron' for their manufacture in each case. I have given my opinion in the answer.

(a) home computer;

(b) a motorway;

(c) an aircraft carrier;

(d) a detached house.

Patronage from the state in one form or another (the community, the king, the church, the military) has been a main basis of major engineering projects. This should not be seen as restricted to ancient times. It still happens. Infrastructural projects such as motorways have generally been state-funded, sometimes for ulterior political reasons rather than simply to supply an

A gallery of images showing a range of impressive buildings can be found on the T173 CD-ROM.

engineering product (witness the German autobahns built by Hitler for the relief of unemployment in the 1930s and the American interstate highways built for the same reason under Roosevelt's New Deal). Perhaps the ultimate example was when US president J. F. Kennedy announced that an American would stand on the Moon by the end of the 1960s. Was he becoming an engineer's client? Not really. He was the spokesman for a *political* purpose of the nation. In those Cold War days it was politically necessary for the USA to be conspicuously ahead of the USSR in space: position in the space race was the marker by which the technological advancement of a country was judged, and it implied military capability. Naturally it was just as important for the USSR to be seen to be ahead. The scientific objective of space exploration was not really the issue, at least for the politicians. Nevertheless the challenge was accepted by the nation as a worthy aspiration, so billions of dollars were invested to make it happen. Thousands of engineers working in teams both within NASA itself and in myriad subcontracted companies strove to produce the vehicles, the fuel and the control, communications and life-support systems which were needed to meet the functional goal.

When and where commerce has grown and wealth has accumulated in hands other than the state's, engineers have found other patrons. Cities such as Venice were built on commercial rather than military success, and ultimately commerce was the basis of the Industrial Revolution. Opportunities for increased markets for goods called for increased production and for greater volumes of raw materials. Hitherto entrepreneurs were the wealthy who speculated their wealth for further gain, but with the scale of investment increasing, the shareholding method of raising finance became ever more the norm. Companies formed for specific ventures contracted engineers to construct the infrastructure for their businesses. The history of engineering has attracted the attention of many scholars, even in ancient times and is so widely described that a few paragraphs here cannot match alternative reading.[1]

After about 1800 we could look in almost any field for examples of consulting engineers: mining, factory production, the growing chemical industry, iron and steel. Their story is as well told as any in the development of railways through the nineteenth century as the major transport system for goods and passengers, at first in Britain and then world-wide. The next section gives a précis of the early history of the railways. Whilst you are reading it, you should ask yourself what was driving this development in engineering: who were the patrons and who were the beneficiaries?

7.3.2 Growth of the railways: engineering for transportation

From at least 1600, horse-drawn wagons in quarries and collieries were often mounted on rails. It is much easier to propel a wheeled truck along a hard surface than a soft one: you will appreciate this if you have ever tried to drive a car across a muddy field. For the rails, cast- or wrought-iron plates of gradually improved design had reduced wear, so the idea of an iron railroad was waiting for exploitation.

The opening of the nineteenth century saw several engineers striving to build steam-powered locomotives, so that the horses could be replaced, leading to faster and more efficient trains. It was realized that by heating water in a sealed cylinder, forming steam at a high pressure, the energy of the burning coal or wood could be turned into movement. Steam will rattle the lid of a saucepan on a cooker as it is driven from the boiling water. A steam engine is merely an efficient way of harnessing this energy contained in the steam.

[1] A book I enjoy and recommend, for its broad historical context, for its insight into the personalities of the engineers and for its interesting choice of illustration is J. M. Parnell's *An Illustrated History of Civil Engineering*, F. Ungar, 1965.

Eventually it was the need of the Durham coalfield to get its coal to the sea which provided the incentive and the capital necessary to beat the problems. The plan was for a horse-drawn trackway and an Act authorising this got through Parliament in 1821 after much opposition. Why opposition? Various reasons, including vested interest of the hauliers who would be done out of business by the transition. By 1821 Edward Pease, the main force behind the project, had met George Stephenson, who had built steam locomotives for colliery use and who persuaded Pease that steam locomotion was the transport of the future. By January 1822 George and his son Robert had submitted a survey of a route from Stockton-on-Tees to Darlington. This was accepted and work began in May. The funding for the project came from public subscription and private enterprise. Three and a half years later the first train, carrying passengers in temporarily fitted coal wagons, covered the 8.5 mile trip, viewed by thousands of spectators. The line was an immediate commercial success both for passengers and goods.

News was not long reaching Manchester and Liverpool, where the merchants and manufacturers in the growing cities were being seriously exploited by the canal owners who monopolized the transport of goods to and from Liverpool docks. Bad service, long delays and high cost were arrogantly provided. 'Why don't we build a railway? It will be faster and cheaper,' the merchants and manufacturers mused. 'Because we shall block the enabling bill in Parliament,' was the canal owners' riposte. (Vested interest again.) Though George Stephenson had been appointed engineer in charge of the project, to get the Bill passed in Parliament required the opinions of other eminent engineers to be sought, and professional conflict soon caused trouble.

From an engineering point of view the project was unprecedentedly difficult. The route included a tunnel, a viaduct, a deep rock cutting and a bog. Also Robert Stephenson, George's son, had gone off to foreign parts and it was soon apparent that his was the organizational skill. The project ran into financial difficulties. A government loan brought in inspectors who found and reported chaos on site. Eventually in 1829 after application for a further loan, the great Thomas Telford, who advised the government, came to meet Stephenson. He did not recommend the loan, mainly on the grounds that the decision about what type of locomotion should be used was too long delayed. What sort of locomotion should be used in this case? The choice was between horses, fixed towing engines to winch the trucks along, and mobile steam locomotives running on the rails.

The railway company planned to settle this issue by competition and set up time trials over 1.5 miles at Rainhill on 6 October 1829. Robert Stephenson, who had returned from his South American adventures, entered his locomotive design prototype *Rocket* for the trials and won the prize of £500. Spurred by his son's success, dad got on with the construction work, to good effect. The line was inaugurated on 15 September 1830 with eight train-loads of VIPs making the trip from Liverpool to Manchester. This was also the scene of the one of the first railway accidents, when William Huskisson MP was run down by *Rocket*. It is worth considering that Huskisson (along with most others at that time) would never before have seen a piece of machinery move as fast as *Rocket*, and had no idea of how little time he had to move out of the way.

Again commercial success followed, and railway fever infected the country. In the next twelve years over 1800 miles of public railway were laid across Britain, from Glasgow in the north to Exeter in the south west. Speculators rushed to invest in any project (rather like the boom in early Internet companies in 1999/2000): the day of privately financed engineering had arrived. Britain's example was quickly followed by the other industrializing nations of Europe and, of course, by the USA. But it was largely British engineering skill and capital that put railways into the rest of the world – the Empire (Canada, India, Australia and Africa), Russia and South America as well.

Exercise 2.2

The Stockton–Darlington and Liverpool–Manchester railways were both built primarily with goods in mind. However, they also became extremely popular for passenger travel. What were the benefits of rail travel to the passenger?

SAQ 2.6 (Learning outcome 2.3)

Think of the modern rail projects that are current in Britain today, for example the Channel Tunnel Rail Link, or upgrades to old lines. Who are the patrons for these engineering projects?

7.4 Engineering for engineers

When we contemplated 'engineering' some shelves, we quickly noticed that ready engineered products would be used: planks, screws and other items from the tool kit. This little example says what I mean by the title of this section: some engineers have done their work so that other engineers can do theirs. The example also shows the three ways in which this idea is manifest:

1 the provision of the materials for doing a job (the pre-sawn wooden planks);

2 finished products that are incorporated into the work (screws);

3 the means for doing the job (your tools).

A great deal of engineering happens in a context where the product, though functional, is moving down a chain of engineering aimed at distant goals. Notice my plural 'goals': planks and screws and carpenters' tools are not made exclusively with shelves in mind. They are versatile products which will find many applications. Here is another, less mundane example.

Making ammonia was a small part of a long chain. Originally Haber and Bosch made ammonia so that other chemical engineers could make explosives, which in turn enabled production engineers to manufacture munitions. All this engineering enabled a military objective to be pursued so that a political end might be achieved – a long and increasingly abstract chain of actions. Haber's and Bosch's research constituted a learning process which eventually determined the starting points for their engineering. The resources they required were a long way back from the ultimate outcome, and not especially close even to the immediate objective of making ammonia. Again the starting points for the pressure vessel fall into the same three categories. They needed an engineered material (weldable steel), engineered products (pumps and power sources), and engineered tools (welding machinery). They did not have to contemplate how to make any of those as part of the chain of actions – any more than scraper-makers had first to make stones. The parallels are exact; to make the functional product requires: (a) knowledge of what's available and (b) knowledge of how to use it.

The functional product of the Haber–Bosch process is ammonia. The engineering was directed at constructing the plant which would produce the product, not at the product itself. Provided the plant is used according to the instruction manual it can't avoid making ammonia. Thus the plant is itself a 'functional product'. This sort of stacking of one function upon another is what we expect of 'tools'. A tool is a functional artefact which can be used to produce another, or play some part in that making. Think of a carpenter using saw, chisel, hammer, plane, screw-driver, etc. each in turn playing its part in producing, of course, a set of shelves.

SAQ 2.7 (Learning outcome 2.7)

So far I have mentioned several finished products which require engineering from other sources in order to enable their manufacture. Examples are soap, a bow and arrow, a ballpoint pen etc. For one or more of these products (or any other product you can see around you at this moment) think of what other forms of engineering are needed in the processes to make them. Decide what are the starting points for the manufacture in terms of previously engineered products in the three categories (materials, products and tools).

Throughout engineering history, the invention and making of tools has itself been a major area of engineering. Often innovation of 'know-how' is represented by engineers making tools with new capability. It has been, and still is, the increasing power, precision and complexity of machine tools, using the word in that general sense of an artefact whose function is to enable another thing to be made, that has brought us so far with our making skills. Ultimately, co-ordinating machine tools dedicated to different functions brings us to the concept of a 'production line', where a sequence of operations by different machines creates a final product. The ammonia-synthesis plant is a chemical production line – different aspects of the production take place in separate parts of the system. The robotic factories for making things such as electronic microcircuits or car bodies are our present 'state of the art' in bringing together, shaping and assembling materials into functional products.

Engineering in its full sense is now the front line of progress. The means of production of functional products depends on the invention and development of machinery. Through much of Europe and the USA, the nineteenth and twentieth centuries saw an explosion of innovation in this area, supported by continuously improving understanding of the underlying sciences; mechanics, thermodynamics, electromagnetism and chemistry. Primary industries such as mining (for coal and for metal ores), the oil industry, the chemical industry (including smelting metals) and the electrical power industry have grown to provide raw materials and energy for other engineers to use as they will. But each of those industries has itself demanded its own engineering. Thus, in addition to the social complexity which characterizes 'civilization', we can now see an engineering complexity characterizing 'industrialization'.

7.5 Engineering for economics

One prime reason for engineering in the civilized, industrialized world, is in order to make a profit. The early railway companies attracted money for their construction from private investors who saw that the railways would become profitable, firstly from transporting goods and subsequently by moving passengers. Railtrack shareholders are the equivalent of those early railway entrepreneurs.

Financial capital is needed to drive industrial development. Whereas an artisan's workshop may be cheap, factories and mines are not. Financing enterprise was initially in the gift of the wealthy, but banks willing to lend for venture have opened the possibilities of being an entrepreneurial engineer. The emergence of new consumer products has often followed this route. Portable transistor radios, personal computers and bikes are examples of successes. The outcome depends on a host of factors: Does the product catch the fancy of the buying market? If it does well, will competitors enter your field? Is their product better or cheaper? What are your costs compared with turnover? Can you expand to meet demand? And so on. Some ventures succeed – the Dyson vacuum cleaner is a recent example at the time of writing. For some, success is to be bought out by a bigger company. Some fail – like the Sinclair C5.

But although innovative entrepreneurship is still practised, this is not the common environment for engineering. Most engineers are *employed* by corporations. As industry creates its own wealth for re-investment, corporations themselves become the centres of wealth and gain control over their own destinies. In *The New Industrial State*, J. K. Galbraith, the acclaimed American economics guru, argues that the *purpose* of corporations is to secure their own survival. You can see one aspect of this in the organization of the multinational corporations. Survival techniques practised at board-room level include taking over competitors and merging so as to become big enough to be invulnerable to takeover. The fashion for 'vertical integration', in which a corporation gets control of all the resources it needs, from its production systems and beyond to its sales outlets, is another organizational ploy directed at self-protection. These are not much to do with the employed engineers, but, if self-survival is the corporate goal, it will influence what engineers are required to do – particularly in two ways.

First there is the matter of investment. A new product, say a new model for a car, needs a lot of money to be spent. There is all the design work for the product which is expensive enough. Not so long ago the biggest expense was to set up the new production facility. If an older production line is being refurbished, it comes out of production (and so is not making money) while this is being done. A brand new line will cost even more – though it may be more productive because it will allow newer technologies to be implemented. All the cost is up front of any revenue from the new product and also has to be seen as risk investment. However good the market research, the product might fail when it comes to the market. Competitors' models might be preferred; fashions might change; in some products (notably in the electronics field), your whole concept might become obsolete during the lead time.

Obviously there is pressure upon the engineers to minimize this investment lead time and its attendant risk. To this end, flexible production lines are becoming the norm, able to produce several models at once. This has been facilitated by computer control. In principle, parts for different models can be organized to arrive at the right place at the right time and the robots doing the welding can be directed to select the appropriate program for the particular car that arrives at a work station. Such flexibility also reduces the risk to the company of a new model that does not prove popular in the market – the company is not committed to a pre-determined and maybe excessive production level. You can see the effect of this if you go to buy a new car these days. If you don't mind waiting a few days, you can specify its details in all respects and the factory will make the car to your order.

Dramatic changes like this do not happen accidentally, but result from the second pre-occupation of corporate engineers – research and development. R&D, as it is always called, should be regarded as a survival strategy for any corporation. Indeed companies and countries which skimp on R&D investment are to be seen as putting their survival at risk. (It is one of the long standing criticisms of post-war Britain that this investment has always lagged behind what competitor countries have been prepared to spend.) So markets disappear as competitors bring innovative products to the market and innovative processes to their production lines.

So where the money comes from can be crucial to the success of an engineering venture. This is dependent on someone being convinced that the project is worthwhile, either in terms of generating profit in the future, or for a government to see sufficient perceived benefit in the project to grace it with funding from public money.

SAQ 2.8 (Learning outcome 2.4)

Think again of the example of the ballpoint pen. In making the pen, it is necessary to have raw materials for all the individual parts, and machinery for turning the raw materials into the finished product.

(a) What do you think are the relative contributions of each of these to the final cost of the pen to the customer?

(b) When is most investment required for producing the pen?

7.6 Summary

At a fundamental level, engineering is something which is part of our creative nature: finding innovative solutions to problems. However, the reasons for grand engineering projects generally go some way beyond this, and arise from the needs of our society. Because these needs are many and varied, there is a variety of instigators and patrons for engineering projects. Crossing a river by a ferry is fine, but might it not be more cost-effective in the long run to build a bridge?

Precisely who makes the choice to instigate a project will vary from country to country and from case to case. However, it's clear that engineering, in the modern world, is something very distinct from craft, and that it's usually possible to work out who is behind an engineering project or a programme of engineering development.

In the following section we'll start to look in more detail at some of the skills which underpin the engineer's activity. I have already hinted at some of these, in talking about innovation and problem solving. For engineering in the modern world, with its range of specialisms and technical complexities, there are many skills which are required by the professional engineer, regardless of whether those skills are applied to electronic device design, mechanical machines, or engineering project management. We will begin to investigate this, along with a look at the intellectual and physical resources which are used.

8 What do we do engineering with?

8.1 Resources for engineering

Let's remind ourselves of our definition of an engineer's task:

> To apply scientific and technical knowledge to the design, creation and use of structures and functional artefacts.

The provision of resources is implicit in this definition. Some resources are found in the natural world – materials and energy – and some resources are found in society – staff and finance. Just as important, however, are those internal resources, the scientific and technical knowledge, that engineers have to bring to the job. In this final section of this introductory block I want to unpack these different kinds of resource to set the scene for all the following studies. The connections with society have been discussed in the previous section. Here, therefore, I shall first consider what internal resources engineers should bring to their work and then conclude the section with reviews of energy and materials resources.

8.2 Who wants to be an engineer?

8.2.1 Generalists and specialists

If being an engineer seems to you daunting, because of the vast range of skills that apparently constitute engineering, don't despair. One problem with this perception of engineering is that its history is littered with stories of brilliant individuals. However, probably none of these brought to the profession every desirable personal quality in abundance: witness George Stephenson's problems when his son Robert briefly abandoned him.

We have already suggested that our species' common characteristics of imagination, creativity and originality are of prime value to any engineer. Some people are more endowed with these than others; but even if you don't have the skills to be an engineering innovator, you should still have the basic skills required to learn, and to apply the skills that you have learnt. Certainly I doubt that you'll have problems in putting up a set of shelves.

The proliferation of engineering into every aspect of production has made specialism essential. Even within a single industry, many specialisms will be found. To build an aircraft requires experts in airframe design, control systems, hydraulics, engines and cabin design (to name a few). Within your chosen specialism, accumulating experience means that you come to each new job with a bit more know-how as your repertoire grows. Beyond that speciality, the experience of working within a team (which is how most engineers work) brings an appreciation of other people's specialisms.

One trouble with experience is its tendency, eventually, to stultify innovation and creativity – as exemplified by the person who knows all the answers but can't see that the questions have changed. It is just as important for an experienced engineer to stay fresh, to be aware of innovations of products, of design, of materials, of processes and of the changing priorities of society. This is probably more important today in the tumult of IT than it has ever been before. So curiosity about everything around us is vital to the engineer who would stay vital.

8.2.2 Communications skills

I have emphasized the organizational and managerial skills of our engineer. Any serious engineering project is now (and probably always has been) of such complexity that it takes teamwork to bring it off. Brunel did not draw the plans for every bit of his railways himself. He was not present at every construction site all the time. He could not supervise the manufacture of rails or the quarrying of ballast. He did not survey every inch of the route, and he could not debate in Parliament for all the land purchases that had to be made. Evidently he was the lynch pin on whom a big team depended for design guidance, for final decisions and for an overall view that would put resources where they were most needed. Boundless energy, immunity from stress, the capacity to keep many balls in the air at once and the ability to communicate accurately and understandably are apparently pre-requisites of the upper echelons of the profession.

But communication is a two-way process, and teamwork involves co-operation among all members of the team. It is not enough that the leader of the team be proficient in these inter-personal skills. Indeed, over recent years, better psychological understanding of group dynamics has led to better structures for teams. Quite apart from reducing 'buck-passing', decisions are best made where there is best information, which is not necessarily at the top of a tree of command. Giving people greater responsibility for their actions generally increases their commitment to the project, and opens routes for creative participation. Communication skills are even more vital in organizational structures where there is extensive team-working.

Language, spoken or written, is by no means the only form of communication used in engineering. Numerical and mathematical expression are inescapable. Hence, the engineer needs to be skilled both in interpreting and presenting numerical, statistical, algebraic and graphical information. Graphs are just one of many forms of pictorial communication you will meet. Other examples are the traditional 'blueprint' (i.e. scale drawings, like that for the ballpoint pen we looked at earlier), system diagrams, dressmaking patterns, electrical circuit diagrams, and so on; no doubt you know of others. Each of these communication devices has its own jargon of symbols which has to be understood before any interpretation is possible.

8.2.3 Knowing what and knowing how

One way or another, in order to *do* engineering you have to be able to solve problems. That is what engineering education is supposed to teach. Problems come in many guises, as we have already discussed – organizational, managerial and financial; plus, of course, all the problems which bear on what engineering is actually to be done, what materials and power are to be deployed and how. It is a great deal easier to say in principle how this is to be done than actually to do it.

One of these 'in principle' descriptions of the process is the so called *modelling cycle*, Figure 2.3. Start at the top where there is a box labelled 'problem'. Now follow the arrow. The arrow leads to a box labelled 'model', and along the arrow the instruction is to create the model. So what do I mean by a 'model'? It's a word with quite a few meanings. In our context a model is a tool for thinking with – an abstraction of the problem, which is what you often need to create to solve engineering problems. Figure 2.3 can itself be viewed as a model of the problem-solving method.

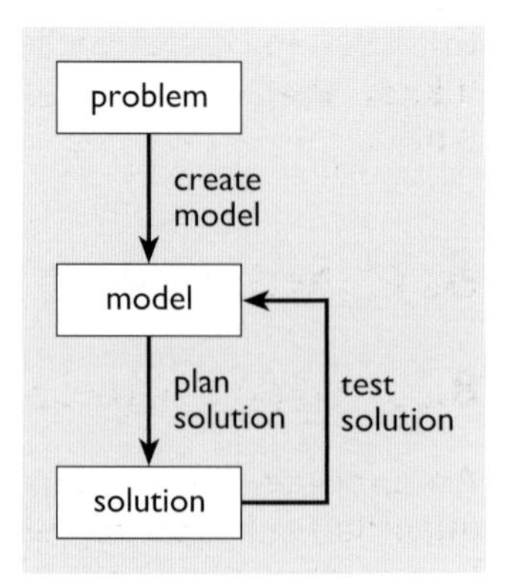

Figure 2.3 A cycle for constructing a model

Going along that first arrow, then, what has had to be created is some abstraction of the problem into a form, the model, that lets you think about it. For example, the engineers who built the Pont du Gard probably created a model of their problem (to get the water across the valley) which was a scale drawing of their proposed bridge. This was a tool that allowed them to think through how much stone had to be quarried, what timber they would need for the centring and what labour to employ. From this example you can see that the next arrow on the modelling cycle tells us to use the model to plan a solution. By this stage you should know what you are going to do.

Or do you? What if your model has not taken all the relevant matters into account? You may have come up with a solution which will not work – you have not really solved the problem. Perhaps, in our example, there are not enough skilled stone masons available. But there is an army of carpenters. Should you re-think the whole project as a wooden aqueduct or look further afield for stone masons? The final arrow on Figure 2.3 asks that the solution be *tested*. Make as sure as you can that the proposed solution actually solves the problem. If there is doubt then go round the cycle again looking for a better model which might produce different solutions. Testing the proposition is a most important step.

Exercise 2.3

Can you think of examples of engineering which have either failed to meet a particular need or, perhaps more often, have created new and unforeseen problems?

I would excuse you if you thought that what I have just told you is self-evident. To solve a problem, it seems obvious that you need to find a way of thinking about it, do the thinking and check that the result. My reason for taking trouble in setting up this model is to think about the *problem* of problem solving. Simply amassing knowledge of engineering principles does not aid creativity. But accumulating know-how from problem solving leads to a repertoire of models for specific, well-defined problems. What makes these models useful is that the methods of thinking with them have also been worked out. The solutions which are produced are going to work; they are good answers to the specific problems.

So, why struggle to 're-invent the wheel'? If the knowledge already exists of how to build an arch or a pressure vessel or an electric motor, why not benefit from it and recycle it. The real snag is that the problems to be solved do not always recur identically. Also, the 'problem' may be couched in very wide terms. How shall we supply water to our city (the citizens of Nîmes)? How shall we beat the blockade of nitrates (Germany in World War I)? How can we demonstrate our technical superiority (Kennedy's moon missions)? How shall we get our coal to the sea (the Durham colliery owners)?

When faced with giant, perhaps quite vague problems, a myriad of potential solutions can present themselves. You won't know which to choose until you can find some way of thinking through the options. Simply resorting to ready-made answers may give a solution that does not really solve the problem at hand. What is easy may not be best, although the client may not appreciate this. There may be a real conflict here, so judging between 'I wonder if …' and 'I know that …' should be part of an engineer's intuitive approach to their work if they are to keep their creativity alive.

8.3 Sources of engineering models

8.3.1 Data sources

An important part in the construction of any engineering model is quantitative knowledge about the physical nature of the problem. This might be something complicated about a material's electronic properties for use in a micro-chip, or as simple as its cost. If polymers were very costly to produce, they wouldn't be used widely to produce the bodies of cheap ballpoint pens.

Data of this kind fill enormous volumes of text. There's far too much to carry in your head, so it's best to know where to find it when you need it. An awareness of where the answer can be found is just as useful as the answer itself. Where to get the information will depend on the problem and its context. The information may consist of tables of data for certain materials, or suppliers' catalogues for specific components. The advent of CD-ROM and DVD technology means that many of these databases are now available in computer-readable form, greatly improving accessibility. So as you set about designing your latest masterpiece, the basis for the necessary calculations can be accessed.

8.3.2 Engineering standards

Engineering standards are another way in which past solutions influence present design. I have mentioned these before in connection with the safety of the high-pressure ammonia-production process. For almost every safety-critical product there is a standard which specifies conditions which must be met during its design, manufacture, testing and service. These standards serve several useful purposes. For example:

- Standards summarize good practice, thus ensuring adequate function and safety of the product. However, although standards encapsulate what has been done before, you should not think of them as inhibitors of innovation. Part of the responsibility of the authorities who administer standards is keeping them up to date and responsive to changes.
- Standards may protect engineers from litigation. An engineering team that can prove their works conform to the relevant standards may have a defence against accusations of malpractice.
- Standards should protect clients from bad engineering. A manufacturer whose products are not fit for function or turn out to be hazardous because standards have not been met can be liable in law. (However, not many standards are enforced legally: some are merely Codes of Practice, rather than strong requirements.)

There is now an International Organization for Standardization (ISO), necessitated by the increasingly global nature of trade in engineering products. But all the major engineering countries have set up standards authorities of their own. In the UK the British Standards Institution issues the standards with the BS code that we exemplified above. In the USA, the American Society for Testing and Materials (ASTM) is in control and Germany has the Deutsches Institut für Normung (DIN). Increasingly, in today's global economy, local standards are being subsumed into international ones.

8.3.3 Scientific models

Nowadays much of what engineers understand derives from science. For example, as we have seen, Haber and Bosch needed to understand principles of chemistry before they could engineer an ammonia plant. On the face of it, these days engineering is often thought of simply as the 'appliance of science'; but, as you appreciate, there was a great deal of engineering success prior to our present, verifiable, scientific understanding of the natural world. None the less, engineering problems often require scientific models to inform their solution.

Moreover, for some phenomena, scientific understanding appears to be remote from a possible engineering application. Knowing about the chemistry of making ammonia will not tell you how much TNT is needed to demolish a building. In this case, even a precise knowledge of relevant scientific theories would not be much use in arriving at a solution. This is not simply because there are random and chaotic processes at the microscopic level that are relevant and which the scientific models are unable to take account of. It is also because scientific models deal with a stripped-down version of reality – one which is sufficiently simplified to allow regularities to be recognized and predicted. It is the method of engineering to live with the reality in totality. The scientific model may therefore be expected to come up against the limits of its truthfulness (or usefulness) before it has described the engineers' domain of work.

The following story may illustrate this. Once I was teaching a class about the scientific understanding of how forces on a particle can make it move along a curved path (see ▼Newton, apples and forces▲). After the class, a trainee from a motor racing team came to me with some disappointment that my lesson had fallen short of where it would be useful to him. His problem, he said, was whether stiffening the rear suspension of the car he was working on would cause it to oversteer or to understeer. The scientific models I had been dealing with, in which the world of moving bodies is restricted to 'particles' (i.e. objects with no relevant internal structure), did not even approach the problem.

▼Newton, apples and forces▲

Figure 2.4 (left) The Pantheon; (above) The Millennium dome

Figure 2.5 An apple, with the downward force exerted by gravity

I mentioned in the first part of the block how weight is actually a force, caused by gravity pulling on the mass of an object. Forces are ubiquitous in engineering. Large structures must withstand the forces arising from their own weight, and a force must be generated to make a car, a train, or a bicycle move forward. Long before Newton such concepts were understood, if not completely rationalized by a scientific law: the Pantheon, built in Rome in 118 BC, has a dome that spanned 43 metres. Compare this to the Millennium dome, built over 2100 years later, and spanning 320 metres.

In 1686 Isaac Newton published *De Principia*, a Latin text which includes a treatise on motion. Newton had thought deeply about the subject, stripping away irrelevancies until the core of the problem became revealed. He realized that there is only one type of force, whether it is the force in the wall of a house from the weight of a bricks or the force generated by a horse pulling a cart.

The *unit* of force is the newton (symbol N), in honour of Sir Isaac. A medium-sized apple, appropriately, is drawn downwards by a force of roughly one newton in the Earth's gravitational field. However, the precise definition of the newton, like that of the second, is not straightforward. More specifically, the newton is defined in terms of motion and mass rather than weight.

Newton realized that forces must always be involved in order to accelerate or decelerate something. A vehicle that has no forces acting on it at all will keep going in a straight line forever. Space probes that are launched beyond the solar system need only have fuel for their initial manoeuvres. Once away from any significant gravitational fields they just travel on without further thrust.

Exercise 2.4

You will be aware that a car will slowly come to a halt, even on a flat road, if its engine is not running. Where do you think the force is coming from to change its speed – to slow it down?

How much effect a force has depends on:

(a) how big the force is,

(b) how big the object is that the force is acting on.

Humans can generate sufficient force to push a car along a flat road, but not enough to push one up a steep hill. The definition of force is coached in these terms: one newton is defined as the force which will give a mass of 1 kg an acceleration of 1 m s^{-2}. Note that we are talking about an *acceleration* rather than a velocity. If the force remains, the object will continue to accelerate. The bigger the force, the larger the acceleration; the bigger the mass, the smaller will be the acceleration if the force is unchanged.

In scientific and engineering writing it's customary to replace words like force and acceleration with a symbol: this is just a form of shorthand writing to keep things tidy.

(The use of symbols is covered in Chapter 5 of the *Sciences Good Study Guide*, particularly Section 2 between pages 119 and 123.)

Force is represented simply by the capital letter F, mass by the letter m and acceleration by the letter a. We can write out mathematically the description of how force is related to the mass and acceleration. Force equals mass multiplied by acceleration, which can be written as:

Force = mass × acceleration

and, using symbols:

$F = m \times a$, or just $F = ma$:

If there is no sign between the m and the a it is assumed that they are being multiplied together.

We now see the advantage of having a common system of units for our measurements. As long as the mass is given in kilograms, and the acceleration in metres per second per second, then multiplying these two together will tell us the force in newtons. If the mass is in ounces, then the number produced for the force isn't meaningful – so stick to SI!

You may realize that 'newton' is just a shorthand way of writing a unit that could alternatively be expressed in terms of kilograms, metres and seconds. Units can be multiplied in the same way as numbers. Mass has units of kg, and acceleration has units of m s^{-2}. So, force could have units of kg m s^{-2}, which indeed it does. The newton (N) is simply a shorthand for this: it makes things tidier and pays homage to a great scientist at the same time.

SAQ 2.9 (Learning outcome 2.4)

If a car is accelerating at 2 m s^{-2}, and has a mass of 1200 kg, what force is acting on it? Where (in simple terms) is the force coming from?

SAQ 2.10 (Learning outcome 2.4)

A ball with a mass of 5 kg is kicked with a force of 50 N. What will its acceleration be?

Gravity generates a force by pulling objects toward the centre of the Earth (or whichever planet you happen to be standing on). Now clearly this force isn't the same for all objects, or everything would have the same weight. What is the same for all objects is how fast they accelerate because of the Earth's gravity. The story goes that this was was first demonstrated by Galileo, who dropped cannon balls of different mass from the Leaning Tower of Pisa: the balls hit the ground at the same time regardless of their mass. The American astronaut Dave Scott also demonstrated this during the Apollo 15 mission to the Moon in 1971. A feather and a hammer dropped together hit the surface of the Moon at the same time.

Why would a feather and a hammer fall at different rates on Earth?

Feathers are designed to maximize their air resistance. The passage of air around the feather generates a force which opposes gravity and so slows it down. Hammers are not as aerodynamically shaped.

On Earth, the acceleration due to gravity is about 10 m s^{-2} (actually 9.81). So (air resistance apart), something dropped from a great height will increase in speed by approximately 10 m s^{-1} every second. After 10 seconds, it will be travelling at 360 km h^{-1}.

Exercise 2.5

You can now calculate how much force is generated by the mass of an object. The mass of a brick is about 2 kg; what is the force due to gravity on the brick? (Use 10 m s^{-2} as the gravitational acceleration.)

The *force* calculated in Exercise 2.5 is the *weight* of the brick, and it is what weighing machines detect. If you could take the brick to the Moon, where downwards acceleration is only 1.7 m s^{-2} (because the mass of the Moon is less than that of the Earth), the brick would weigh only 3.4 N. It would still have a *mass* of 2 kg.

The example of dropping the feather carries a very useful message, which is that we have to consider *all* the forces which are present before we can do a calculation. The feather has an upward force from air resistance as well as a downward force from its weight, so its acceleration is reduced. A mug sitting on a table has a downward force from its weight, but it doesn't accelerate downwards because the table produces an upwards reaction force.

The understanding of forces is critical to how structures work and how vehicles move. Neither the Millennium Dome nor the Pantheon could be constructed without some understanding of forces and of the need not to exceed the strength limits of the materials used. The Dome is a much larger structure than the Pantheon, made possible by modern materials and modern methods which allow us to calculate accurately what the forces will be in each part of the structure.

Forces are defined in terms of acceleration, but buildings are normally required to stay still. The trick in a building is to ensure that forces are always balanced, that there is no force in any direction which is not opposed. The designer also needs to ensure that external forces, from wind for example, can be opposed by forces in the walls which won't exceed their strength.

8.4 Resources of energy and power

In the old times, when horses pulled barges along canals, there was a particular straight and narrow section on the Duke of Bridgewater's canal where the horses knew how to take life easy. They discovered that, if they got their barge up to a certain speed, its bow wave reflected off the banks, caught the stern of the boat and drove it down the canal. For a little way Dobbin could trot down the tow path with the rope virtually slack. He did not have to *work* so hard; that is, he did not use so much *energy*.

The concept of energy lies at the very heart of engineering and science. Like mass, length and time it is something which is easy to appreciate but hard to define. One definition is that 'energy is the capacity for doing work'. This is perhaps too abstract. Another view is that energy is something which is transferred between parts of a system when something happens. So on the canal, energy was transferred from the horse to the barge and thence to the water waves. (The trick that Dobbin mastered was to go fast enough that the waves in the canal worked to push the barge forward, rather than just lapping away uselessly.)

Energy is what you buy on your electricity bill and is transferred from the mains to the water in your kettle (see ▼Units for energy▲). To put the electrical energy into our homes, energy held in a fuel (like coal or gas) is transferred by burning the fuel to make heat which raises steam in the power-station boiler. Energy in the steam is transferred to a turbine and via a rotating shaft to a generator, thence down the line to the kettle.

Since it is the engineer's job to make things happen, understanding and controlling these transfers of energy is important. The energy we use derives from fuel (horse feed could be viewed as a fuel), or from a device engineered to extract it from wind, water or enriched uranium. In practice, then, energy costs money, even if the source is free (as with sunlight and wind power). Hence it is important to transfer energy efficiently (that is, with minimum waste). Inventing, designing and building 'engines' – energy transfer machines – has occupied a great deal of engineering endeavour over the ages. Dobbin is

▼Units for energy▲

Dobbin can help us find some units for energy. The horse walks along the tow path exerting a *force* on the rope to tow the barge. At the end of the day he is tired because he has gone a long distance. He has done a lot of *work*, and used a lot of *energy*.

Scientifically, work is defined as the product of the force exerted and the distance moved. So the units of work, which is one form of energy, are newtons × metres (N m as symbols). This time the honoured scientist is Joule who did a lot of experiments in the field of energy transfer. A joule of work, or energy, is defined as 1 N m, and is given the symbol J.

A joule can be expressed in the base SI units in the following way:

$$\begin{aligned} 1\ \text{J} &= 1\ \text{N m} \\ &= 1\ \text{kg m s}^{-2} \times \text{m} \\ &= 1\ \text{kg m}^2\ \text{s}^{-2} \end{aligned}$$

Once again, having a shorthand for the unit (J) looks like a sensible idea.

There are many different types of energy. Dobbin was converting the chemical energy in his feed to energy of motion of the boat and the water. Energy possessed of a body by virtue of its motion is called 'kinetic' energy ('kinetic' comes from the Greek word κινεο = to move). You see that the 'work' done by the horse has been converted into kinetic energy of the barge. Other forms of energy, electrical, thermal or chemical will still be joules. Unlike velocity, acceleration and force, which are all vectors, work and energy are not vectors.

SAQ 2.11 (Learning outcome 2.4)

Dobbin pulls a barge with a steady force of 0.2 kN for a distance of 20 km. How much work has been done? Give your answer in megajoules.

▼Units for power▲

How quickly a machine or horse can generate or transfer energy is related to its *power*. Power is the rate at which work is done. Indeed, one of the earliest units of power was the 'horsepower' defined as 550 foot pounds weight per second. That is a force (the weight of a mass one pound) *times* a distance (one foot) *per* unit of time (one second).

The SI unit of power is defined as one joule per second, and is known as a watt (symbol W, named this time after James Watt, pioneer of the development of the modern steam engine – the stationary type, some time before they were adopted as mobile units for the railways). For interest, a horsepower is about 750 W. So

$$1 \text{ W} = 1 \text{ J s}^{-1} = 1 \text{ kg m}^2 \text{ s}^{-2} \text{ s}^{-1} = 1 \text{ kg m}^2 \text{ s}^{-3}$$

The watt is a lot simpler than the long-winded version of the units!

SAQ 2.12 (Learning outcome 2.4)

If Dobbin did the 20 km in the last SAQ in 5 hr 33 min 20 seconds, at what power was he working?

If you are confused about the combination of terms for these units, you might want to review the Maths Help section on Powers and Roots, pp. 353–355 of the *Sciences Good Study Guide*.

one answer. The ubiquitous windmills and water mills of the Middle Ages represent further engineering progress. Development of steam engines through the eighteenth century was the enabling technology for the Industrial Revolution.

What is the nature of the 'progress' that has taken us from animal power to steam engines? Neither is very efficient; much energy does not get converted into useful work. But the steam engine delivers a given amount of work in a much shorter time. A big strong horse will tow the barge faster than a little one – it will do the same amount of work in a shorter time. A bigger horse, or better still a steam engine, is more *powerful* (see **▼Units for power▲**): it delivers the energy more quickly.

In the late eighteenth century, James Watt greatly increased the efficiency of his steam engines (see **▼Less than 100%▲**) by providing a condenser to condense the steam back to water as it left the cylinder. This extracts extra heat energy from the steam. In 1774 he set up in business with Matthew Boulton, and their engines were greatly prized for their power and reliability. Ironically it was Watt who defined the horsepower, to describe how many horses his engines would do the work of. They could always do the work of several: the chemical energy released by burning coal could be converted into mechanical work at a far greater rate than any other available technology provided. A workshop full of textile machines (spinning jennys, looms etc.) could be powered from a single engine driving a shaft from which power could be tapped off by belts running on pulleys (not a system that would meet today's safety-at-work standards).

We have come a long way since Watt, inventing many different types of engine using various fuels: internal combustion engines, like petrol and diesel engines for cars, and gas turbines for aircraft; rocket engines using controlled chemical reactions; and electric motors. We have learned to design engines for

▼Less than 100%▲

In talking about converting energy, whether from coal to electricity at a power station, or electricity to light in a light bulb, we're often interested in how *efficient* the process is. A person or a horse probably only uses 25% of the energy which is available potentially from the food they eat. The same is true for my other examples: not all of the energy available in the coal will end up as electricity, and not all the electricity will be turned into light energy. There may be many reasons why processes are less than 100% efficient, but essentially it's very difficult to harness all the energy from something in a useful form. There may be energy lost in overcoming frictional forces, or in generating heat that's wasted.

virtually any power we choose and know how to select what is appropriate for a given task.

Table 2.1 Power ratings of typical devices

Device	*Power*
TV remote control	1 W
car side light	10 W
room light	100 W
kettle	1 kW
central heating boiler	10 kW
sports car engine	100 kW
earth-moving machine	1 MW
railway locomotive	5 MW
engine of large ship	100 MW
large power station	1 GW

It is worth getting some idea of the power output or consumption of various 'engines'. Horses are stronger than people, so you should guess a power of rather less than 750 W for a human. A typical, slightly fit person can sustain an output of 100 W on a bicycle using the big leg muscles, whereas a super-fit sprinter can output perhaps 300 W over 100 m. At a smaller scale, I calculate that if a flea weighing 0.1 g jumps 10 cm high, reaching the top of the jump in a tenth of a second, that is a milliwatt or 10^{-3} W, which is enough to run a megabyte or more of computer memory these days. Table 2.1 gives you a handle on greater powers.

Perhaps a Boulton and Watt engine would have delivered a few tens of kilowatts. The power station in the table at 1 GW is a steam engine some hundred thousand times more powerful. Its function is the same as Boulton and Watt's – to be a central power source serving distributed applications. Using electrical distribution rather than shafting and belts, it is a good deal easier to spread the power over longer distances and to apply it to many different functions.

I dare say that the power station is also a good deal more efficient than Watt's engines, but it is still limited to less than 40% efficiency. Nineteenth century scientists strove to understand what governed the efficiency of steam engines, developing the science of *thermodynamics*, which looks at how useful energy can be harnessed from heat. It soon became apparent, at least to French physicist Sadi Carnot, that there is a maximum possible efficiency for any conversion of energy from heat to any other form. Unfortunately even this 'best possible' only applies if the energy conversion is infinitesimally slow, that is, if it happens at zero power. In contrast, conversion from electrical energy to other forms is usually very much more efficient, greater than 90%, say, for a well-designed motor. Ways have been sought to get energy directly from fuel, without going via heat energy contained in high pressure steam, but without conspicuous success. So called 'fuel cells', which make the direct conversion, are currently restricted in power and in choice of fuel.

8.5 Material resources for engineering

8.5.1 Product, property, process and price

The outputs of engineering are always **products** of some sort, either, as we have seen, directly functional for a consumer (cars, TV sets) or steps in a chain of engineering towards a wide range of products (tools, chemicals). Either way they have to be sold into a market, and, in competitively structured economies that means the selling **price** is important. Nowadays, engineering products include software for computers, but apart from such 'intellectual' products, everything has to be made of some material (and indeed software requires the hardware of a computer in order to be of any use). To meet the proposed function of the product, whatever it is, the materials are chosen for their particular **properties** (a bow has to be springy; a floor beam has to be stiff). To make the shape which the product requires involves **processing** the material (the bow has to be whittled to symmetry from a branch and notched to hold its string). These four words, product, price, property and process, are intimately linked (Figure 2.6). By examining a few examples of the linkages we shall be able to see how, in the past, the changing availability of materials governed what engineering could be done, and how, in our present times, the vast repertoire of materials influences design choices.

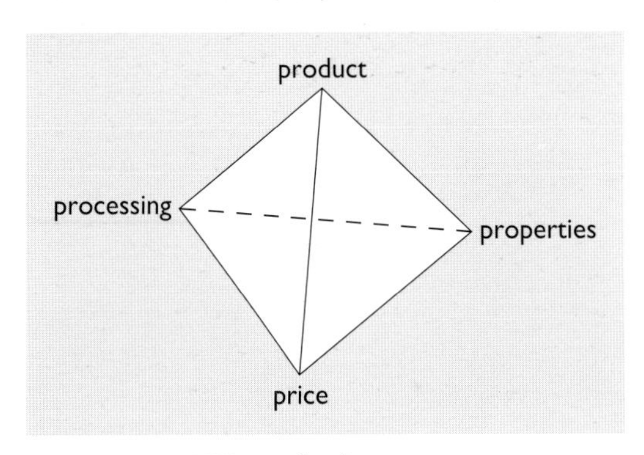

Figure 2.6 PPPP tetrahedron

8.5.2 Ancient materials

Archaeology has recognized the critical significance of materials in the names given to successive stages of our engineering development – Stone Age, Bronze Age, Iron Age. To an extent these are artificial conceptions of the times, dictated by what has survived for archaeologists to find. Of course, many more materials than stones were used in the Stone Age, but very few organic artefacts (wood, bone, leather) have survived to be found and catalogued by the archaeologist.

Nor was everything made of bronze in the Bronze Age. But with the discovery of bronze, new engineering horizons were opened and those innovations mark a cultural shift for the archaeologists. And, whereas copper and tin have inconveniently dispersed sources, so bronze-making is sometimes out of the question, iron ore is common everywhere. Discovering how to win iron dramatically increased the opportunities for making things of metal. The point is that, generally, the range of materials was added to. Only occasionally would an earlier material go out of use (for example, stone axes). As an example of ancient material diversity look at **▼Travel gear for the late Stone Age▲**.

Already I have hinted at one major step in materials knowledge, the possibility of artificial materials. In the very beginning, humans, like chimpanzees, used only naturally occurring materials. Stones and wood, obviously, but also animal hides, hair, grease and bone, and other vegetable products, like bark, leaves and fibres.

▼Travel gear for the late Stone Age▲

Five thousand years ago, a man died whilst crossing the Alps between Italy and Austria. His body and the things he was carrying (Figure 2.7) were found in 1990.

The 'Iceman' wore leather clothes and shoes, had a woven grass cloak and had stuffed his shoes with grass. Technically, the implications of these are that, in his culture, people knew how to make a durable material by tanning animal hides, were acquainted with weaving methods and were aware of the thermal insulation qualities of cellular materials.

His axe had a copper head slotted into a wooden haft and lashed with a leather thong. The copper was imported, so there was trade with other places as well as a knowledge of metalworking. His axe haft and his bow were both made of yew wood, which has springy qualities useful for both applications. A twisted twine of grass fibres was used for the bow string, so spinning fibres into a strong yarn was part of his technology.

He carried a knife whose flint blade was set in an ash-wood half, implying that the ability of flint to be worked (knapped) to provide sharp cutting edges was known, as were the qualities of ash that made it easier to shape than yew. He also had a carrier or pouch made of birch bark, which provides a stiff sheet material.

By modern standards, these materials technologies may seem modest, but it is clear that each material had been deliberately selected from those available to suit its intended function.

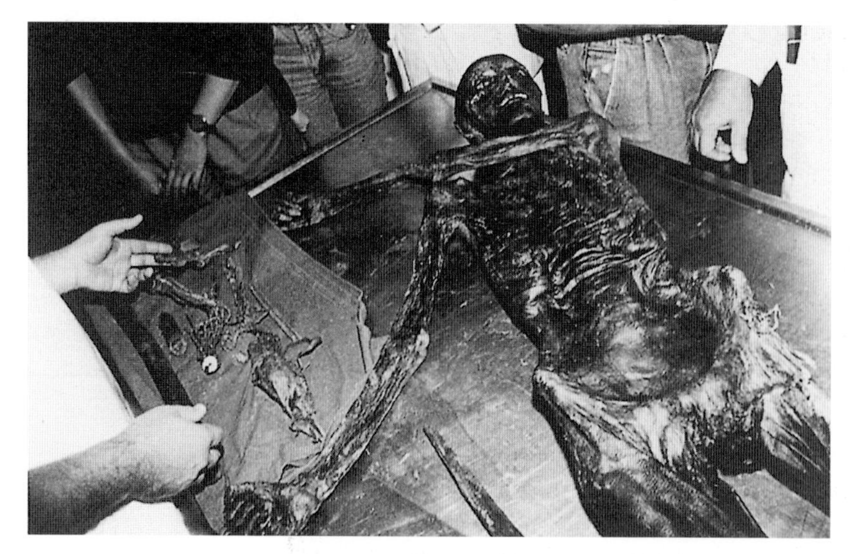

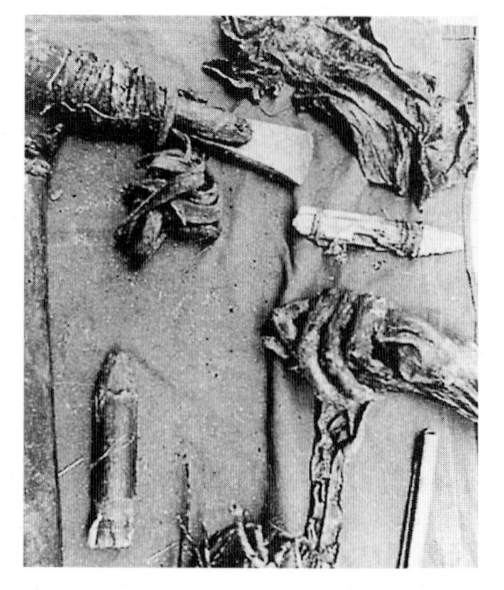

Figure 2.7 The 'Iceman' and his belongings

SAQ 2.13 (Learning outcome 2.7)

Use your imagination to list the processes that would turn stones, bone, hide and wood into functional artefacts.

All these processes accept the properties of the materials as given. Shapes are changed (by cutting etc.) and parts are assembled (twisting threads from staple fibre), but no attempts are revealed which seek to change the properties of the materials themselves. Probably the very first *processing* of a material to change its properties was hardening wood by fire, perhaps for a more penetrating spear point. Another quite early discovery might have been that of tanning hides to leather by boiling the hide with bark. Leather has an altered property: it does not rot so fast as raw-hide.

Pottery is the first durable processed material to appear in the archaeological evidence. Here is an interesting advance: the shape of the object is made while the material is in a soft, workable condition and then a separate firing process changes the property to hard and stiff. Considering the difficulties inherent in shaping stone, this route is effective in providing 'artificial' stones of chosen shape and size, in the form of bricks and tiles for building.

The earliest fired clay objects recovered were small figurines (fertility idols?) modelled by hand and, presumably, fired in the campfire. Hollow pots as vessels were, of course, not much use to nomads; they would have been both heavy and brittle, neither of which is particularly useful for people who are continually on the hoof. Functional artefacts such as pots and bowls came a surprisingly long time after settlement. Both the shaping process and the firing call for ingenuity, but techniques for handling clay successfully developed in every civilization. From a materials resource point of view, potting is not just a matter of digging up the local clay. Clay is a very variable commodity, and knowing what is best and how to use it is all part of the skill.

Probably potting was one of the first specialisms to separate craftsmanship from intuitive engineering, but it is worth noting that production often reached industrial scale in ancient times, implying the development of infrastructures and the need for organizational skills. Thus the division is not always sharp. For example, the zenith of technical achievement in ceramic art was reached with Chinese porcelain in the early part of the last millennium. Porcelain requires the mixture of two ingredients, known in Chinese as *kao-lin* (white china clay) and *pe-tun-tse* (a fusible rock, actually itself a natural mixture of quartz and feldspar). Both these ingredients had to be finely ground and refined by water sedimentation before they were usable, so there had to be a materials processing industry to back up the potters' needs.

Before long, potters found that the hotter the fire the better the pot, and kilns with air draught came into use. To build kilns able to contain fire at over 1300 °C and large enough for industrial scale production, as at the Chinese porcelain town of Ching te Chen, calls for real engineering.

Getting high temperatures by using an air draught was also vital to the next advance in materials availability: winning metals from ores. Tiny quantities of gold and silver can be found as metal from the sand in streams – if you know where to look. Occasionally copper can be found as metal, just needing to be melted from the rock. The Damara tribe in Namibia knew of such a place on the edge of the Namib Desert and made a living trading copper jewellery. But to get respectable amounts of metal you have to disentangle it from chemical compounds. Essentially, then, at this stage we began to make artificial materials, ones that scarcely exist in nature. Depending on which metal and what compound it is in, the process is more or less easy. One of the ores of copper is the mineral malachite, in which the copper is combined with carbon, oxygen and hydrogen in various compound forms. At a temperature readily reached in a crucible within a fire blown by bellows, this mineral will

▼Pours for thought▲

One problem with metal is how to process it into a shape once you've extracted it from an ore. Metals have some super properties: they are strong and tough. But these very properties make it difficult to work them into a usable shape. Without complex tooling they are hard to cut and require enormous force, in Bronze or Iron Age terms, in order to squeeze them into new shapes. The solution is to melt the metal: it can then be poured into a suitably shaped mould.

The Iceman (Figure 2.7) carried a copper axe. To make this, the first step is to make a wooden 'pattern'. Round this a mould is formed of clay in two parts. When the clay has dried off, the two parts are separated and the wooden pattern removed. Put the mould back together and pour in molten metal. Let it cool, open the mould and there you have the axe shape in copper. If the mould remains intact, it can be re-used.

react with more carbon (such as the charcoal used to heat the fire) to release the molten metal. How are we to make use of a lump of copper? It did not take long to invent a new process – casting (see ▼Pours for thought▲).

Copper on its own is not well suited for making cutting tools because it is quite soft. Compared to flint (say), its market competitor as a material for an axe, copper's advantage lies in the process for making it; casting is much quicker than knapping, and a better shape can be formed for fixing the head to the handle. Before long, however, the metal material was improved by alloying tin into the copper to make bronze. This harder metal would keep its edge quite well. Wherever bronze was available, flint was displaced from the market. Notice that this is another artificial material, an invention to improve a functional **property**. By 1500 BC the Chinese had added lead into bronze for a **processing** improvement. A fashion **product** was cast bronze vessels with intricate surface patterns; the extra ingredient made the molten alloy much runnier, ensuring that it filled fine details of the mould.

8.5.3 Iron

The other great metallurgical breakthrough of ancient times was to extract iron from iron ore. The advantage of iron over copper alloys is the commonness of iron ore, but to get it presents a much greater technical challenge. Higher temperatures are needed, the carbon used for the chemical process of extraction dissolves in the iron affecting its properties, and not all the ore reacts, leaving slag within the metal. Hence the initial result of the extraction process would often be a dull, brittle metal, not much more use than the ore from which it had been extracted. The next process was therefore prolonged hammering and repeated folding of the hot metal to burn out the carbon and to disperse the non-metal slag finely through the material. This is *wrought iron*, which is tough and strong; it was the best engineering material for the next two thousand years (see ▼Iron and steel▲).

The developments that turn wrought iron into steel were a long time coming. To turn wrought iron into steel you have to get close control over the amount of carbon in the metal and, except for a few isolated instances of very small-scale production, that was a step that had to wait until well after the start of the Industrial Revolution. But expanded production of iron was a necessary precursor of that revolution and the seventeenth and eighteenth centuries saw

▼Iron and steel▲

Steel is one of the most ubiquitous materials in the modern world: it is used for cars, construction, packaging, casings, and generally finds employment anywhere that a cheap metal will do the job. Steel is an alloy of the element iron. It always contains some carbon, which has a great strengthening effect in iron, and other elements are often added to the mix to produce certain particular properties: nickel and chromium, for example, are added to provide corrosion resistance for applications where rusting needs to be avoided.

Oddly, in terms of the nomenclature, steel is a purer form of iron than the 'iron' which is obtained from the initial smelting of the ore.

a great expansion in the use of iron. Timber supplies for charcoal were being depleted, and could not sustain ever-increasing iron production. It became necessary to use abundant coal (refined to coke by heating without air, just as charcoal is made from wood). But that had its problems: we now know that it was the sulphur in coal that ruined the iron quality. Abraham Darby solved the problem fortuitously in 1709 at Coalbrookdale, north west of Birmingham, where the coal happened to be low in sulphur. By the 1760s, when the Industrial Revolution is reckoned to be getting going, most iron was already smelted using coke.

Two other innovations helped in the efficient production of good quality iron: adding limestone to the charge was discovered to put all the non-metallic parts of the ore into an easily melted slag, so the iron could be cleaner; and then the invention of the *blast furnace* which could be loaded continuously with iron ore, charcoal and limestone with the molten metal and slag run out from the base.

The iron which comes from blast furnaces is easily castable, and has a lower melting temperature than pure iron, because it contains about 4% by mass of carbon which dissolves into the iron during the smelting process. Because carbon is much less dense than iron this figure is deceptive; it represents more like 20% by volume (see ▼Mass and density▲). When the iron solidifies virtually all the carbon forms sheets of weak carbon between the crystals of iron. Figure 2.8 shows a specimen examined under a microscope.

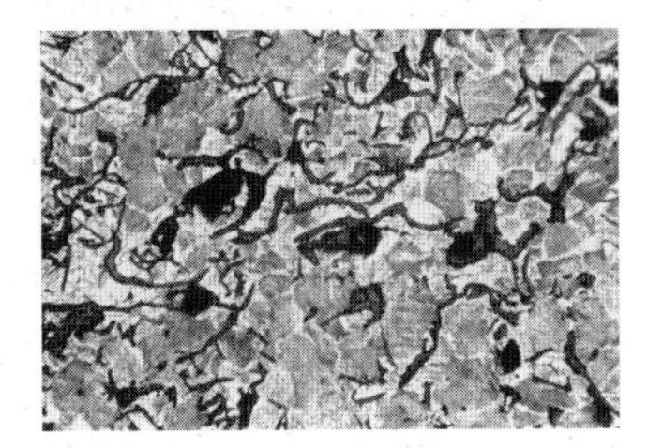

Figure 2.8 A microscope picture of a cast iron containing carbon flakes, which make the material brittle. The black regions are the carbon flakes, the lighter area is the iron. Magnification approximately 100×

In consequence, cast iron is a brittle material, almost useless in tension. But as it became available in bulk, it could be used as a constructional material if it was kept in compression (remember this concept from the previous part of this block). That was just the same problem as with stone, so you find designs in cast iron which are reminiscent of stone. Look at the photograph of the cast iron bridge at Coalbrookdale, Figure 2.9, designed with arches to ensure compressive stresses.

The engineers building canals, railways, steam engines, bridges, docks and ships used cast iron enthusiastically. Just as the ancients had to get the carbon out by laborious hammering to make stronger wrought iron, so now the search was to do the same trick in bulk. Henry Cort developed the 'puddle furnace'

▼Mass and density▲

You are doubtless aware that the same volume of different materials does not necessarily mean an equivalent mass. A big bag of mushrooms will weigh much less (and so, has a lower mass) than a similarly-sized bag of potatoes. This is because the mushrooms are less *dense*.

A dense material is one that has a large mass for a given volume. The SI definition of density is that it is equal to the mass in kilograms of 1 cubic metre of material. Finding the density of a material from a sample is easy; simply divide its mass by its volume:

Density (in kg m^{-3}) = mass (in kg) ÷ volume (in cubic metres)

The symbol given to represent density is actually a Greek letter: ρ (rho, pronounced to rhyme with 'toe'). Science, mathematics and engineering make extensive use of the Greek alphabet to supply extra symbols beyond the 26 available from our alphabet.

The use of symbols is covered in Chapter 5 of the *Sciences Good Study Guide*, particularly Section 2 between pages 119 and 123.

I have already noted that mass is given the symbol m. Volume has the symbol V.

So in symbols $\rho = m/V$ and the unit is kg m^{-3}.

SAQ 2.14 (Learning outcome 2.4)

A lump of carbon has a volume of 0.2 m^3, and a mass of 500 kg. What is its density?

SAQ 2.15 (Learning outcome 2.4)

Iron has a density of 7860 kg m^{-3}. (Note that this is quite weighty: one cubic metre has a mass of nearly 8 tonnes.) What will be the mass of a strip measuring 250 mm × 15 mm × 2 mm? (*Hint*: you will have to calculate the volume first, and remember that you will have to work in cubic metres.)

Figure 2.9 The cast iron bridge at Coalbrookdale

Figure 2.10 Telford's Menai suspension bridge. The high span was dictated by Admiralty insistence that navigation of the Menai Strait should not be impeded.

where cast iron was melted, separated from the coke-burning fire, and stirred to expose it all to air. The carbon burnt off, leaving malleable iron. For the first time engineers had a material in bulk quantity which could carry large tensile stresses. Look what that allowed. Figure 2.10 shows Thomas Telford's masterpiece of a suspension bridge to carry the main road to Holyhead onto Anglesey. It was opened in 1826. Of course the catenary chains and the suspending rods of such a bridge are all in tension.

8.5.4 Modern times

I won't trace the development of the whole portfolio of materials accessed before our scientific age. To the metals mentioned so far we have to add only zinc, mercury and lead. Non-metallic artificial materials in addition to ceramics (both bricks and pottery) included glass and cement. Compared to our present range it was a meagre selection and did not grow much from Roman times. Indeed, if Julius Caesar, who conquered France in the first century BC, had returned to visit his soul-mate Napoleon Bonaparte some eighteen hundred years later, he would not have found many new materials (though I guess he would have been impressed by gunpowder). The materials boom came from then onwards, both from the growing understanding of chemistry and by applying the experimental approach to process development with more control and determination.

Exercise 2.6

Look around your home or workplace and make two lists of artefacts under the headings of 'ancient' and 'modern' materials. Try to get at least half a dozen items on each list.

Nowadays, to a much greater extent than ever before, because of the vast choice they face, engineers have to know the capabilities of their material resources (see ▼What is a 'material property'?▲). Choosing the right material both for manufacturing convenience as well as for functional performance, to say nothing of getting the price right, is not a simple or trivial matter. How this can be done will occupy your attention deeper into the course. But first let us begin to understand how the materials themselves can be described.

The list you made in Exercise 2.6 will have indicated the diversity of new materials, and even then you will not have seen or recognized many that are

What is a 'material property'?

The range of materials properties is vast, from strength and stiffness, which we judge instinctively before sitting on a chair or climbing on a table, to electrical and magnetic properties that are non-intuitive and relatively obscure. In modern engineering, quantification of such properties, by measurement and tabulation, is essential in order for correct selection to be made of a material for a specific application.

An important point to consider is that *materials* properties are different from the final properties of *components* which are made from them. There will be an element of geometry: how much material is used and what form it is in. A good example of this is shown in Figure 2.11: one of the bookshelves shown is sagging dangerously, even though both shelves are made from the same material.

Clearly one of the shelves is too thin. One solution is to change to a thicker shelf, or use a stiffer material; or supply an extra support in the middle.

The physical properties of an object are always some combination of its *material properties* and its *geometry*.

Figure 2.11 Bookshelves

hidden from view, such as the cunning ceramic in the detector of security lights which senses the heat from a passing body, or the materials which coat the inside of your TV screen and glow with different colours. We cannot review all those, so to conclude this section I will say a few words about two types of materials that should have been on your list – aluminium and polymers.

Aluminium

Essentially, right up to the twentieth century, there were just the two structural metals: copper (and its alloys) and iron (including steel) which came to us from long ago. Now, wherever you look you find aluminium: in saucepans, kitchen foil, beer cans, window frames, power cables, and so on. Most importantly, aluminium is the stuff for building aircraft; it is also finding use in some lightweight car bodies. It is a lovely metal: light, strong, non-corroding, and easily formed and machined. It can be hardened by alloying with a little copper. It melts at a lower temperature than the 'old' metals, so processing it doesn't have to be at too high temperature, keeping energy costs down. It can even be given fancy colours though the process of anodizing. Moreover it is even more abundant than iron.

So why did the ancients not recruit it into their materials repertoire? The answer lies in its chemistry. Whatever aluminium combines with, it hangs onto with great tenacity. It is impossible to separate the metal from its ore by reacting with charcoal at high temperature (as can be done with iron), and that was really the only metal-smelting process available till recently. The route to aluminium is *via* electricity, and that we only began to use in the nineteenth century: a commercial process to win aluminium electrolytically still waited for the twentieth century. Independently, Hall in the USA and Héroult in France discovered how to make a low-temperature melt of aluminium minerals which could conduct electricity to yield up the metal. A mixture of the minerals cryolite and fluorspar could dissolve aluminium oxide and melt at 880 °C. Pass thousands of amps of current at a low voltage through the liquid and aluminium was released from its chemical bondage. Again, it was the process that was all important. Figure 2.12 sketches the engineering. Even so it is expensive chemistry; it takes 14 MW h of electricity

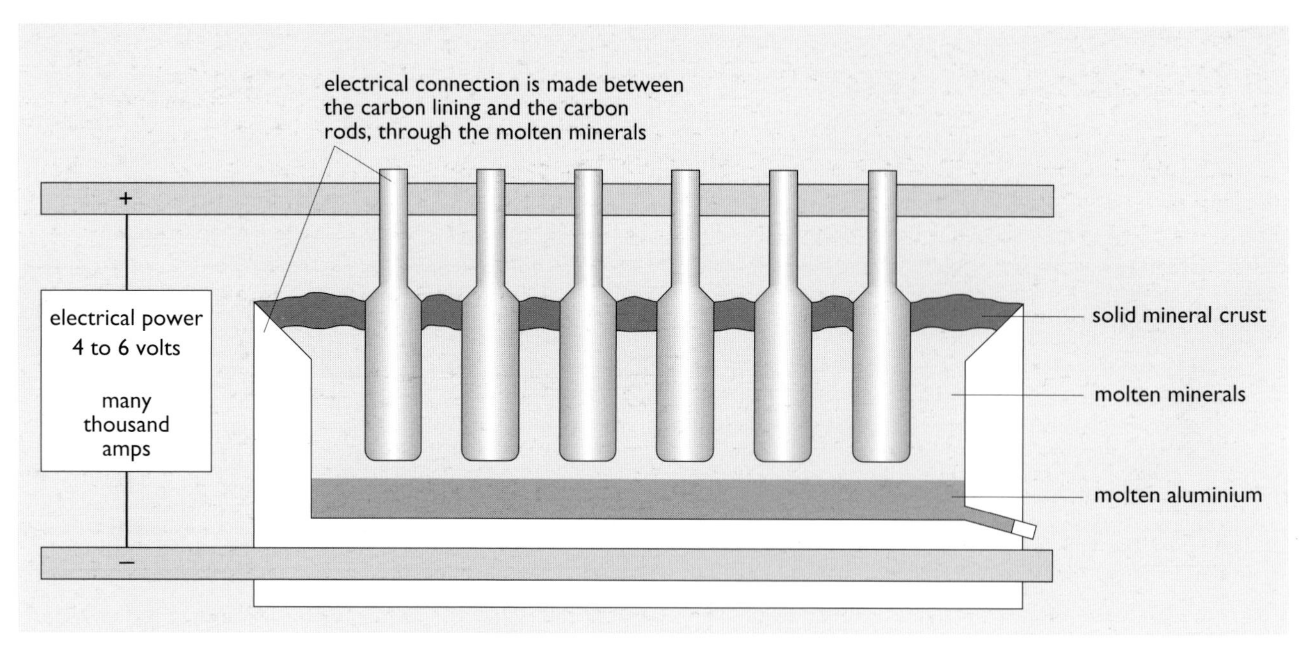

Figure 2.12 Equipment for the Hall-Héroult process of aluminium extraction

(see ▼Time is money▲) to produce a tonne of metal. That is nearly £1000-worth of electricity at domestic electricity prices, so aluminium smelting has been concentrated where cheap hydroelectric power can be produced.

I have mentioned aluminium and steel as though each was a single material. The whole truth is much more complex: there are dozens of alloys of aluminium and even more varieties of steel, each designed to have a distinct set of properties. Metallurgy is a science of its own and it is well understood how to design alloys to bring out particular virtues. And of course the bulk metals go beyond the iron, copper and aluminium mentioned here so far. For the hottest parts of jet engines you need nickel alloys. At the cooler end (though still at temperatures too high for aluminium to withstand), titanium alloys find favour, again for low weight. The filaments in electric light bulbs are tungsten and are supported by molybdenum wires. It goes on. Of the ninety or so naturally occurring chemical elements, some seventy are metals and probably at least fifty of these have found their way into metallurgists' recipe books.

▼Time is money▲

When you receive a bill for electricity or gas, you may notice that what you are billed for is not energy in J, or power in kW, but kW h (kilowatt hours). The bill could be expressed in joules or kilojoules, but this is thought less easy to relate to your usage of appliances.

All the appliances in your home consume energy at a certain rate, so they have a power rating given in watts (or more usually kilowatts). If you have a kettle, it will have a power consumption of around 1 or 2 kW. A 1 kW device running continuously for one hour uses 1 kW h of electrical energy, which costs about 6 p. A 100 watt bulb running for 10 hours also uses 1 kW h.

SAQ 2.16 (Learning outcome 2.5)

Remembering that 1 watt is defined as a rate of energy use of one joule per second, calculate how many joules of energy are equivalent to 1 kW h.

The commercial price of aluminium is currently around £800/tonne, so compared with steel at £100/tonne it looks a good deal more expensive, suggesting that there have to be strong reasons either in performance or manufacture to prefer aluminium to steel. In some applications, such as cladding sheets for buildings where the stress level is low, you are really buying a *volume* of materials so, because aluminium is only a third as dense as steel (2580 kg m^{-3} compared to 7860 kg m^{-3}), the price difference becomes a factor of just over three, rather than a factor of eight. But when it comes to load-bearing, aluminium alloys are neither as strong nor as stiff as steel, so the design may require larger members in aluminium to carry the load. The density advantage won't then benefit the price as much, but you'll usually end up with a lighter structure. That's why aluminium alloys are the right materials for aircraft; less weight means more payload or longer range – or both – for the same fuel consumption.

Polymers

If we are to regard the modern panoply of metals as products of chemical ingenuity then how much more so is it with plastics. Indeed you can really only understand what plastics are by appreciating the chemistry of their construction, albeit at a simple level. Once again we are not talking of one material but a whole class. You can see that in your home; the washing-up bowl is different from the TV box and both are different from the disposable foam plastic cup or cling film wrapping. **▼Of shoes and ships and sealing wax, and cabbages and kings▲** therefore introduces sufficient chemistry ideas for you to grasp what plastics are.

Plastics are very long carbon chains with various bits hanging off the sides. If you can manage to string together somewhere between several hundred to a few thousand carbon atoms in each chain, the result will be a *polymer*, another Greek-derived word meaning 'many pieces'. Some polymers are used pure as plastics, but as with most working materials, many plastics in use are mixtures. 'Vinyl' paints, for example, contain polymer and a white filler (titanium oxide) plus other colourings.

▼Of shoes and ships and sealing wax, and cabbages and kings▲

Democritus, an Iron Age philosopher in Greece, pondered the constitution of matter. Was it continuous, or discrete? He imagined cutting a piece of stuff in half and half, again and again, and wondered if you would ever come to a piece so small that a further cut would mean that those two bits were no longer the same stuff. He coined the word *atom*, meaning 'not cuttable' for that ultimately small particle of a substance. Some 2500 years later the evidence of scientific research confirmed that this was indeed the case. All matter is made of discrete particles. The chemists have found that there are different sorts of atoms. As mentioned in the first part of this block, a substance in which the atoms are all of the same sort in terms of their chemical behaviour is known as an *element*. Ninety-two of these are to be found in nature, and nuclear scientists have made another dozen or so in minute quantities. We now know that all atoms are themselves built of three types of even smaller particles – protons, neutrons and electrons – but we need not go into the details of that here. Also we know how small atoms are, of the order of 10^{-10} m (one tenth of one billionth of a metre). The tiniest speck you can see with your naked eye is a million atoms wide.

Chemists give each element a symbol, which is another type of scientific shorthand. The symbols are either a single capital letter or two letters, one capital and one lowercase. Some are fairly straightforward, like O for oxygen, H for hydrogen, Al for aluminium; others are more obscure, either owing to the fact that the names of some of the elements have evolved over the centuries from Greek and Latin sources, or because the element has a symbol related to its discoverer. So for example iron has the symbol Fe (from the Latin for iron, *ferrum*). (You will not be required to remember the symbols for the elements.)

More information on chemical symbols is given in the *Sciences Good Study Guide*, p.128. There is also information on how chemical symbols can describe chemical reactions between elements.

Chemistry is the study of the patterns of how different elements behave, and how they combine; there are many patterns that can be established which help us to understand the behaviour of the elements. Chemistry can impinge on engineering in many ways, from the extraction of metals, to the corrosion of metals (essentially extraction in reverse), to the requirements to manufacture chemicals for industrial processes.

One of the ways in which atoms combine is to form molecules. A greatly simplifying concept in understanding how molecules can be formed is to think of atoms as having a certain quite well-defined number of 'hooks'. (In reality there are no such hooks, but to give a more realistic account would take us beyond the scope of this course.) In general a stable molecule (which is a group of atoms) can form if all the hooks on each atom are linked to hooks on other atoms in the molecule. Thus with water, as each hydrogen atom has one hook, and each oxygen atom has two, the rule can be obeyed by hooking two hydrogen atoms onto one oxygen atom. We can make a little picture of this by showing the linked hooks as dashes, like this: H–O–H.

That is all the theory you need to know in order to understand quite a lot about the molecular structure of different plastics, and from those structures we can infer quite a lot about the physical properties of plastics.

Table 2.2 shows a few elements by name, symbol and valency (which is the proper name for the hook number).

Table 2.2

Element	*Symbol*	*Valency*
Carbon	C	4
Nitrogen	N	3
Oxygen	O	2
Hydrogen	H	1
Chlorine	Cl	1

The simplest of all plastics is that shown in Figure 2.13, a long, long string of carbon with hydrogen atoms linked onto the sides of the chain. String is a good image, for these long molecules are flexible, and when the solid forms they can fold and organize themselves into crystals. This happens at many places during solidification with several molecules engaged in each organized region and reaching from one to another, tying them together. This plastic is polyethylene (often called polythene), the stuff of supermarket bags. You will recognize it as a flexible, strong, tough and lightweight material when you think of filling a bag with food tins. As a thicker block of material, and with special attention paid to making ultra long molecules, this material serves as the bearing surface in artificial joints for hips and knees. No ancient material could do that job when the original material wore out.

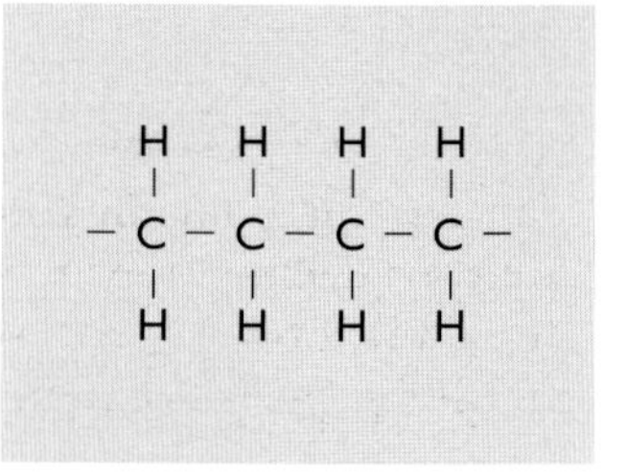

Figure 2.13 Structure of polyethylene. The lines between the carbon (C) and hydrogen (H) atoms indicate that they are bonded (joined) together chemically.

If anything other than hydrogen is on the side of the chains, the flexibility of the chain is reduced and the properties change. Figure 2.14 shows some options (the long dash at each end of the chain fragments is to imply repetition of the same fragment many, many times).

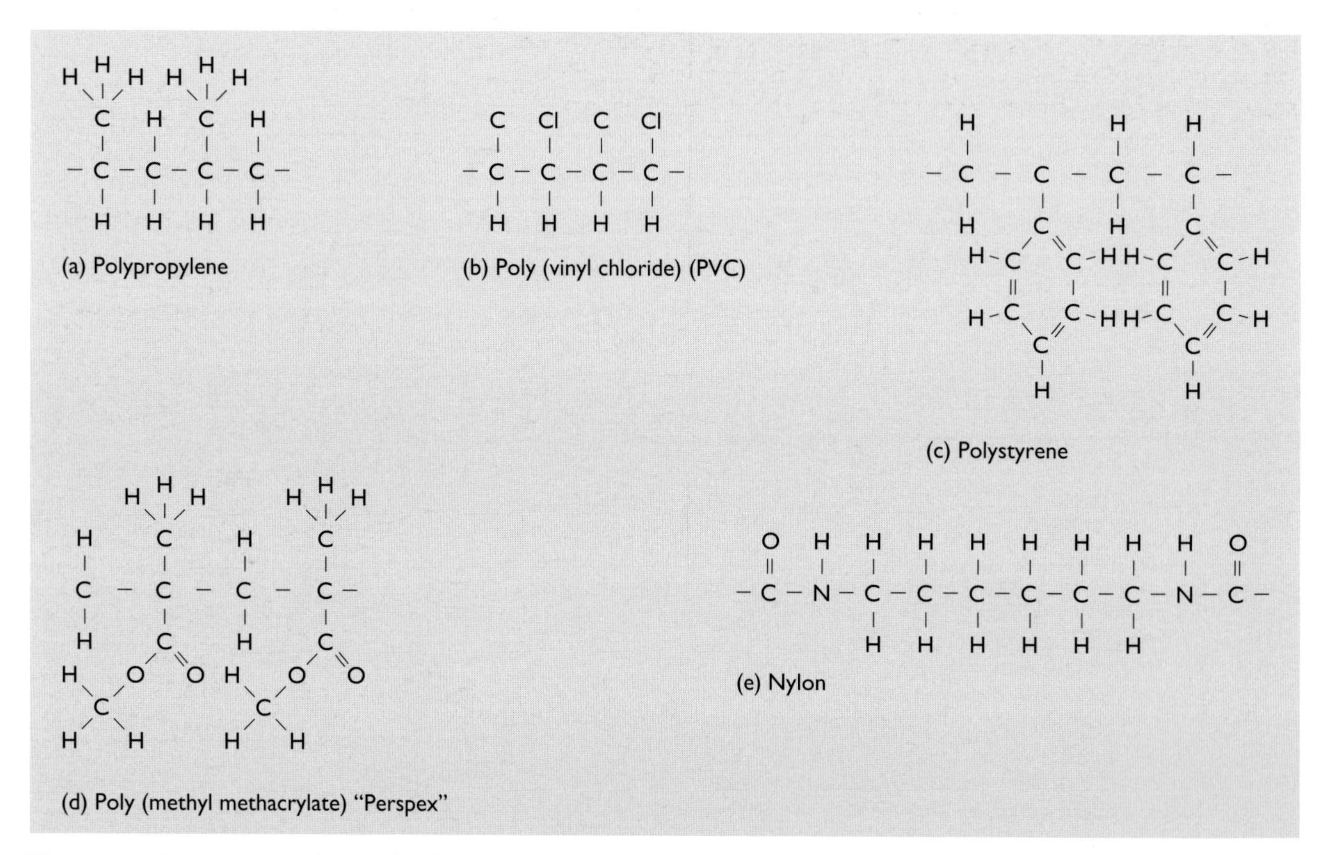

Figure 2.14 Repeat units of several polymers with C–C backbones, and one other. Cl is chlorine, and N is nitrogen. Where there are two lines joining atoms, this shows a stronger bond than usual

The side groups inhibit regular folding of the strings and, for the bulkier groups (e.g. Figure 2.14(b), (c) and (d)) there is no tendency to crystallize at all during solidification. Polystyrene is used for clear stiff food trays, and PMMA is 'Perspex' ('Plexiglass' in the USA) used for applications as diverse as aircraft windows and wallpaper paste. Figure 2.14(e) is nylon and shows that not all our plastics have –C–C– chains.

All the examples I have given are thermoplastics, which means that they can be easily melted, and then processed in modern variations on casting. That opens a wonderland of cheap, fast processing routes. It is no surprise that plastics have such a strong niche in our product markets. A great variety of polymers is produced in the thousands of tonnes by the petrochemical industry. I have given you only a glimpse into the world of plastics; we shall return to them later in the course.

SAQ 2.17 (Learning outcomes 2.2 & 2.7)

You should recall from Exercise 2.6 that you were able to identify a wide range of materials in your surroundings. Doubtless many of these were plastics. Taking one or two examples, what materials might have been used to do the job if plastics were not available? How might this impact on the product from a functional viewpoint and its cost to the consumer?

8.6 Summary

The Industrial Revolution heralded the return of engineering into the business of artefact production, taking it away from hand-crafting. Not only the consumer products but also the processes of production – power sources, materials and machinery – became the subjects of engineering. This, together with continually increasing scientific understanding of the physical world, has provided ever-widening scope for imagination and creativity. The present gigantic scale of the engineered world, both in terms of the quantity and variety of products is the result. Now, as the biological and information sciences are exploding, we stand at a new threshold. Where will it take us?

9 Whither engineering?

Prediction is difficult, especially about the future.

Attributed to Samuel Goldwyn

Quite near the beginning of this block we discovered the key differences which separate humans from other creatures. We can design and create things for purpose, and have done so apparently since we crossed the 'threshold to humanity'. In consequence we expect to be able to *predict*: the inventor of the wheel predicted its behaviour in the mind's eye before the thing was made. Even before the motions of the Earth, Moon and Sun were correctly described, their patterns had been spotted and eclipses were being predicted. And now, with our effective scientific models to aid our engineering designs, how can we avoid believing that everything around us is predictable?

Fortunately for our sanity and for our free will it is not like that. In almost every *detailed* regard, things are unpredictable (the only certainties are taxes and death, as Benjamin Franklin said). Although popular fancy would have it otherwise, even science itself recognizes this. At the heart of modern physics lie principles which deny detailed predictability in the Universe.

Chaos theory is a burgeoning branch of mathematics finding application, among other things, in weather forecasting. At a session on the Global Atmospheric Research Programme at a meeting of the American Association for the Advancement of Science in 1972 Edward Lorenz[2] delivered a talk entitled *Predictability: Does the flap of a butterfly's wings in Brazil set off a tornado in Texas?* to explore how the concepts of chaos relate to the behaviour of the atmosphere. But the catchy title has been suborned into the mechanical view of science – you will see the idea of chaos reported in the media as 'the flap of a butterfly's wings in Brazil causes a tornado in Texas' with a causal connection from the minute to the majestic to rob chaos of its true beauty.

Moving into human affairs, in which I would include the forthcoming changes that engineering might wreak in our world, unpredictability is even as Samuel Goldwyn allegedly teaches. But, undaunted, the advent of the computer fuelled a 'Strong Artificial Intelligence' movement, peaking in the 1980s. Its protagonists maintained that the computer is a good metaphor for a brain, even to the extent that consciousness would be possible once a large enough computer was stocked with a large enough input of knowledge in the form of rules. That was not technically possible so a weaker AI perspective led to an enormous investment by government and industry in 'Expert Systems'. Put simply, Expert Systems are collections of rules that can be processed in a computer. Perhaps the axioms and precedents of law can be formalized within a computer? A research project was funded to model the British Nationality Act.

> The hope was that a piece of legislation could be represented as a set of axioms whose logical consequences could then be tested by means of a computer-based theorem prover. Our use of logic to formalise legislation is based on the thesis that much of human knowledge and belief can be formulated as propositions with a clear logical content.

This work was criticised by Philip Leith, in a 1987 *Guardian* article, who observed that 'the real essence of law is not embodied in the written legislation, but rather in the machinations through which different groups within the overall legal process negotiate meaning …' So the workings of the law depend on the fertile imaginations of these negotiators, not on rule-following of any sort. As we said earlier, one of the nice things about imagination is the surprises it springs. Lawyers have not been put out of business by computers over the intervening decade.

[2] This talk is reprinted in Lorenz (1995).

What has all this philosophical preamble got to do with my predictions for the future of engineering? Is it just an apologia in case I get it wrong? No, the point is that the particular route by which we have come to here was as unpredictable as the lawyers' negotiations, the effects of the butterfly or the quirks of quantum physics and evolution. There was no logic by which the next step followed the last. That moment when the wheel formed in a mind could equally well have been elsewhere, at some other time and in another mind – or not at all: it did not occur to the Amerindians. Then, of course, the whole 'progress' of engineering would have gone a different way. So now, how can I forecast what intensely significant fragment of imagination is going to happen in the next ten seconds or minutes or years? Precisely because imagination, creativity and originality are the essence of what comes next, it is not predictable by any rule or model or logic, and never has been. One rule is to rule nothing out – a very eminent scientist in the later years of the nineteenth century stated very definitely that 'heavier than air' flight was impossible. Fortunately the Wright brothers did not know of this.

However, what I can do is to ponder the contexts in which people's creativity will operate. That might suggest some fields for curiosity. But note that it is only me doing the pondering – you can do your own, or argue in the pub about it; you are no more likely to be right or wrong.

The environment

The environment is a fashionable context. There are even glimmerings of international cooperation in 'doing something about it'. Global warming, believed to be consequent upon our use of fossil fuels having released too much carbon dioxide into the atmosphere, is the major worry. Computer simulations based on current evidence point to ocean levels rising by some hundreds of metres. That would give us something to think about as much of the agricultural land would be inundated. What might be the engineering fix? Very high sea walls? Undersea food production systems? Colonize another planet? What is interesting about our response so far is that the technologically advanced societies of the world cannot suddenly give up making CO_2 because we now have a society which is completely locked into continuous provision of goods and services through trade. Access is governed by having money to join into the markets, so the means of getting money are of primary importance to individuals, communities, corporations and countries. That demands power and fuel. So a context for engineering of almost immediate need is to replace our present power supply systems with solar generating methods. The science is already well understood but the economics favour pumping oil and digging coal, but when push comes to shove...

Exercise 2.7

Present power demands from all sources are about 10^{12} W. Suppose we think we need ten times this amount eventually and that a solar power plant is only 10% efficient. Earth presents an area of about 10^{15} m^2 to the Sun and receives power as sunlight at about 1 kW m^{-2}. What fraction of the Earth's surface would need to be dedicated to power generation if only solar power was used as a source of energy?

The exercise gives an interesting answer. It is not entirely unthinkable – perhaps it's a better bet than building those sea walls. If electricity were used to break down water into oxygen and hydrogen the latter could (with problems) be a portable fuel – entirely unpolluting as it just burns back to the water it came from. Maybe we can keep our beloved cars.

I have speculated at some length on just one perceived problem. It implies a responsible social attitude on a global scale to make anything like my conjecture happen. Within the political context of my world I don't have great faith; maybe we shall come up with other fixes. Might bio-engineering with genetic modification of food crops have a place in the same vein of environmental changes requiring responses? Shall we be looking for much greater yields from our reduced land? Could we be looking instead for ways of reducing population – we are pretty good at that. Or shall we just run as lemmings towards Armageddon, content with e-commerce, instant entertainment, computerized stock markets and the devil take the hindmost?

In the remainder of the course, as we begin to look in greater detail at particular aspects of engineering, we will always try to keep an eye on possible developments in the near future. Of course, any such speculation is informed ultimately by what we know today, so to a large extent prediction in engineering often just assumes that we will better exploit existing technologies. This does help to keep our feet on the ground: there is no point in assuming that a new wonder technology will arrive to solve our future needs.

But that's enough pessimism: let's look on the bright side.

Small is beautiful

I have mentioned information technology (IT) as one of the consuming interests of the present day. Several factors have made the current revolution in computing possible, but a principal one is the miniaturization that is possible with modern microelectronics. In modern microprocessors, the circuitry is engineered at a scale of 0.1 micrometres, rather smaller than can be seen under the best optical microscope. But even that might be considered coarse. At the research level, the ability exists to place single atoms where you want them. So, if I look into the future of microelectronics, I don't see the limit of what is possible too close just yet.

Miniaturization is not limited to the electronics field either. Very small mechanical parts, such as tiny gears, can be made by machining materials with a plasma beam. The prospect of very small 'mechatronic' machines may be closer than we think.

Minutely into space

Then there is the exploration of space. The orbiting Hubble telescope represents one way of doing it. Another is the use of autonomous vehicles, to Mars, Venus, and the distant planets. But in all these applications we are still hampered by the mere 1% payload that results from having to escape Earth's gravity. Either we need a more efficient and cheaper method of reaching space, or what we send up must perform more function for less mass. My speculation for the future is one of micro-spacecraft of prodigious capability, being sent afar by relatively low-powered rockets. Towards the end of the course we shall have more detailed discussions of these possibilities.

Closer to home, there are reasons why we might want to continue space engineering. Perhaps the most commercially impelling reason is the need for communications satellites. There are many such satellites, providing our global communications, and at the current rate of growth in communications there is demand for more.

At lower orbital altitudes, satellites can be used to survey land use, pollution, weather, and resources. They can also, of course, be used for espionage. Again, many have been launched and many more will be.

The other aspect of space engineering that is relevant to our course is to do with the organization of such projects. How to manage projects of great complexity is as important for engineers as being able to design the hardware. This is an issue we shall pick up in the next block.

10 Learning outcomes

When you have studied this part of the block you should:

2.1 Be able to distinguish between craft and engineering.

2.2 Realize that engineering is a progressive activity, and that developments occur continually.

2.3 Be able to identify the patron, social driver or customer for an engineering development or an engineered product.

2.4 Understand that, for large engineering projects or the beginnings of a new engineering venture, that finance or capital expenditure is required which may contribute to the cost of the product for many years.

2.5 Be able to recognize and work with SI units of velocity, acceleration, density, force, work and power.

2.6 Be able to perform simple calculations using numerical data in SI units.

2.7 Realize that engineered products require starting materials, tools, and products from other engineering processes in their construction.

Answers to exercises

Exercise 2.1

Metres and seconds, respectively.

Exercise 2.2

Faster travel than by horse-and-cart or canal. More comfort. Presumably cheaper, as for the early railways passengers were just extra profit?

Exercise 2.3

Some engineering examples which failed to meet the need they were designed for were the first three attempts to put a lighthouse on the Eddystone rock just outside Plymouth harbour. The first two attempts completely under-estimated the power of the seas in winter storms and were swept off the rock. The third try caught fire because the materials around the candles which provided the light were not fire resistant.

An engineering solution which generated new problems was the Aswan Dam on the Nile. Because the flow of the river is controlled, the annual dumping of fertile silt in the delta has been stopped, greatly impairing the agriculture of an area which has been a bread basket for Egypt for thousands of years. Of course the silt is now dumped in Lake Nasser, behind the dam, reducing its capacity unexpectedly quickly. The dam has apparently not turned out to be a good solution to the problem of supplying Egypt with electrical power.

There are many other examples. Nuclear power can supply cheap electricity, but with the problem of generating highly toxic waste material. And so on.

Exercise 2.4

There are forces associated with friction between the wheels and the road, and in the axles and other moving parts of the car. These act to decelerate the car.

Exercise 2.5

$$\begin{aligned}\text{Force} &= \text{mass} \times \text{acceleration}\\ &= 2\ \text{kg} \times 10\ \text{m s}^{-2}\\ &= 20\ \text{kg m s}^{-2}\ \text{or}\ 20\ \text{N}\end{aligned}$$

Exercise 2.6

Table 2.3 is a selection of materials found in my study. As you will have found, we still use the ancient materials but have added a host of new ones, and new processes.

Table 2.3

Ancient	*Modern*
Wood (door, desk)	Plastics and polymers (lots of them, lots of applications)
Paper (fairly ancient)	Stainless steel (fountain-pen body)
Cotton (clothing)	Nickel alloy (various coins)
Pottery (cup, mug)	Several batteries (what are the materials there?)
Glass (window, paperweight)	
Leather (wallet)	
Brass letter opener	

Exercise 2.7

The Sun is providing $1000 \times 10^{15} = 10^{18}$ W to the Earth's surface and we are projecting demand at 10^{13} W. At 10% efficiency we need to collect 10^{14} W of sunlight, i.e. 1/10 000 of the Earth's surface. That is an area 130 km × 130 km entirely coated with solar panels. A chunk of the Sahara Desert would do the job nicely!

Answers to self-assessment questions

SAQ 2.1

I imagine that you're sitting in a room where there are chairs and a desk: depending on your preference for furniture, these could be mass-produced items (with lots of production engineering involved in their manufacture), or hand-crafted. You may have ornaments and so on which are the product of craft skills. Anything more advanced, which requires interaction between many people and complex manufacturing systems, comes under the heading of engineering. This might be a television, a light bulb, or a gas fire.

SAQ 2.2

Of course, the answer depends on how old you are. I would imagine that for most readers of this, the list could include compact discs, mobile telephones, laser-guided weapons, portable computers and printers, and MRI body scanners. For an older generation, it could include air transport, contact lenses, prosthetic hip joints, and many pharmaceutical drugs.

SAQ 2.3

A time of 2 minutes is 120 seconds, and a distance of 2 km is 2000 m. So the speed is

$$(2000 \div 120)\ \text{m s}^{-1} = 16.7\ \text{m s}^{-1}$$

Note the importance of the units: without them the answer to the calculation is meaningless.

SAQ 2.4

(a) The change in speed is

$30\ \text{m s}^{-1} - 15\ \text{m s}^{-1} = 15\ \text{m s}^{-1}$

The magnitude of the acceleration is the change in speed divided by the time, which is therefore

$15\ \text{m s}^{-1} \div 8\ \text{s} \approx 1.9\ \text{m s}^{-2}$

(The answer is actually $1.875\ \text{m s}^{-2}$, but this is cumbersome, and is also gives more *significant* figures than were available in the original numbers for the problem. For more details about rounding and significant figures, see Section 6 of the Maths Help in the *Sciences Good Study Guide*, from p. 346.)

(b) The average speed is $800\ \text{m} \div 150\ \text{s}$, or $5.3\ \text{m s}^{-1}$. (Note that this is the average speed. The athlete starts from an initial speed of zero, and during the race almost certainly varies the speed around the average value.)

SAQ 2.5

The home computer is (usually) made by a company. The 'patrons' who initially stump up the money are the company shareholders. They do so hoping that a consumer will buy the product for more than the company paid to make it.

A motorway is usually funded by the state (or if not, is monitored by state control), to enable the populace to get about more easily.

An aircraft carrier is paid for by the military (a branch of the state concerned with its defence).

A detached house is probably funded by a construction firm, who hope to sell it on at a profit to a customer. Alternatively, someone may decide to hire a firm of builders to construct a house to a unique design.

SAQ 2.6

For large infrastructure projects, like railways and motorways, the finance generally comes at least partly from the state. However, many rail developments in Britain are now privately funded to a greater or lesser degree; as when railways were first developed, companies hope that by constructing and running a railway, a profit can be made over and above the costs incurred in the building.

SAQ 2.7

I'll take the example of the ballpoint pen for this answer. The pen requires the following starting materials: brass for the metal top part (which requires metal extraction from ore); a polymer for the main body (from the polymer production industry, using chemical feedstocks); the tungsten carbide ball (I wonder how that's made); and the ink (again, from another production route). The metal piece will be machined using specialized tooling, all of which has to be engineered. The polymer body will require some sort of polymer casting process, probably with liquid polymer being injected into a mould. This will itself be a piece of complex engineered machinery.

The principles are the same even for a primitive bow and arrow, although the technology is not as advanced. The materials involved are all natural, but they have to be gathered. Some sort of tool is required for whittling the arrow, and this might be a stone tool which is manufactured by a knapping process. The 'string' for the bow could be a piece of animal hide that is cured and then cut to the right shape.

SAQ 2.8

(a) The machinery will be used to make tens of millions of pens. So although the initial cost of the machinery will be large the cost per pen is small. The raw material cost is likely to have a larger contribution to the cost paid by the customer.

(b) Most investment is needed right at the start, to purchase the equipment with which to make the pens, and raw material for the initial batches. There may also be investment needed in marketing etc. to bring the product to market successfully. Of course, the company may not have the capital to pay for the machinery in one go. It might have to arrange a bank loan or some other form of finance to afford the investment. So the cost of the machinery may be spread over five or ten years. With interest to pay, this could double the actual cost of the equipment. The cost per pen is still relatively small, however.

SAQ 2.9

Using $F = ma$, the force is equal to:

$$1200\ \text{kg} \times 2\ \text{m s}^{-2} = 2400\ \text{N}$$

Remembering our prefix of k for 1000, we could write this alternatively as 2.4 kN (said 'kilo-newtons').

In simple terms, the force is being generated by the engine.

SAQ 2.10

In this case we know the mass and the force:

$$50\ \text{N} = 5\ \text{kg} \times a$$

So by switching this round:

$$a = 50\ \text{N} \div 5\ \text{kg}$$
$$= 10\ \text{m s}^{-2}$$

SAQ 2.11

$$\text{Work} = \text{force} \times \text{distance} = 0.2 \times 10^3 \text{ N} \times 20 \times 10^3 \text{ m}$$

Notice the two terms 10^3 which represent the kilo prefix in kN and km. In this case, then, the work done is

$$4 \times 10^6 \text{ J} = 4 \text{ MJ}$$

This is quite a lot of energy for an animal to expend. A man doing heavy manual labour would use probably only 4 MJ in a day's work and would feel tired from doing it. Of this expenditure only some 25% would be output as mechanical work, the rest is used in moving body parts and lost as heat by sweating. The horse's 4 MJ is the actual output energy.

SAQ 2.12

You have to convert the time to seconds. That's

$$5 \times 3600 \text{ s} + 33 \times 60 \text{ s} + 20 \text{ s} = 20\,000 \text{ s}$$

The work done was 4×10^6 J, so the power is 200 W.

SAQ 2.13

Stones were bashed with each other to get shape. With skill this became the craft of 'knapping', producing sharp flakes for cutting tools. These cutters could be used for felling trees, cutting branches, stripping bark, whittling bits to shape and drilling holes. Bone would be subject to similar processes to make needles or barbed spear points. Hide could be cut into strips for thongs. Hair or vegetable fibres could be twisted to form thread or cord. Stitching pieces of hide was presumably an early craft. Shelters could be built of lashed poles and covered with leaves or hides. Archaeological evidence for virtually all of these ideas can be found; we can only conjecture how far earlier humans got with the stone tools we have found.

SAQ 2.14

Density is mass/volume, so

$$\rho = 500 \text{ kg}/0.2 \text{ m}^3 = 2500 \text{ kg m}^{-3}$$

SAQ 2.15

The volume of the strip is

$$(250 \times 10^{-3} \text{ m}) \times (15 \times 10^{-3} \text{ m}) \times (2 \times 10^{-3} \text{ m}) = 7.5 \times 10^{-6} \text{ m}^3$$

Rearranging the equation for the density, we see that:

$$m = \rho \times V$$

So the mass is

$$7860 \text{ kg m}^{-3} \times 7.5 \times 10^{-6} \text{ m}^3 \approx 0.059 \text{ kg or about 60 g.}$$

In this answer, I've changed from mm to m using 'powers-of-ten' notation. One thousand is $10 \times 10 \times 10 = 10^3$, so 250 mm can also be written as 250×10^{-3} m.

These different representations are used both because they are easier to write, and also because it's much better to convert to and work in the correct SI units early in a calculation. The metre is often quite a large scale compared to what is being measured, hence the need for using the powers of ten notation. If you need to know more about representing small or big numbers this way, have a look at the *Sciences Good Study Guide*, pp. 140–1.

SAQ 2.16

A power of 1 watt is 1 J s^{-1}, so 1 kW is 1000 J s^{-1}, or 1 kJ s^{-1}. This means that every second, a 1 kW appliance uses 1 kJ of energy. In 1 hour there are 3600 seconds. In this time, 3600 kJ of energy will be used. So 1 kW h is 3600 kJ, or 3.6 MJ.

SAQ 2.17

I'll take two examples: plastic casings for things like computers and televisions, and ballpoint pens.

Early televisions had wooden cabinets. The development of plastics meant that these have been almost completely replaced. Plastics are much cheaper than wood, and are also cheaper to manufacture, as they can be moulded into shape easily: more manual labour input is required for wooden cabinets. The plastics may also be of lower density than the wood, allowing the cases to be made lighter. So there is a functional advantage in terms of low weight, and a cost benefit to the consumer. There may also be aesthetic considerations, depending on your interior décor!

For the ballpoint pen, the main advantage is cost. The cheapest metal-bodied ballpoint pen I found with a quick look in the shops was ten times the price of the cheapest plastic-bodied version. Wooden-bodied pens were even more costly. So this is an example of where the low cost and ease of manufacture of polymers has allowed a great reduction in cost. The pen doesn't have to be disposable: there were some refillable ballpoints with plastic bodies that were more expensive (at least ten times), where the extra cost arises because they contain more parts (and are aimed at a different market).

References

Galbraith, J. K. (1967) The New Industrial State, Bobbs-Merrill (second edition Penguin books 1972 and 1991).

Lorenz, E. (1995) The Essence of Chaos, UCL Press.

Acknowledgements

Grateful acknowledgement is made to the following sources for permission to reproduce material in Part 1 of this block.

Part 1

Figure 1.1: © Ann Ronan Picture Library/Image Select; *Figure 1.2:* Courtesy of The Africat Foundation, Namibia; *Figure 1.4:* © National Motor Museum, Beaulieu; *Figure 1.5:* © Ann Ronan Picture Library/Image Select; *Figure 1.6:* © Ann Ronan Picture Library/Image Select; Figure 1.9: Reproduced by permission of Allan Jones; *Figure 1.13:* Reproduced by permission of Rachael Dawson; *Figure 1.15:* © Royal Armouries; *Figure 1.16:* Reproduced by permission of Don Eigler, IBM; *Figure 1.20:* © Lynette Cook/Science Photo Library.

Part 2

Figure 2.1: © Telegraph Colour Library; *Figure 2.2:* © National Motor Museum, Beaulieu; *Figure 2.4(left):* Architectural Association Photo Library/Gardner/Halls; *Figure 2.4(right):* Architectural Association Photo Library/Richard Rogers; *Figure 2.7:* © Rex Features Limited; *Figure 2.9:* © Martin Bond/Science Photo Library; *Figure 2.10:* © Architectural Association Photo Library.